LIGHT ELEMENTS

LIGHT ELEMENTS

□ ■ □ ■ □ ■ □ ■ □ ■ □ ■ □ ■ □

Essays in Science from Gravity to Levity

□ ■ □

Judith Stone

Ballantine Books ▪ New York

 Published in the United States by Ballantine Books, a division of Random House, Inc., New York, and simultaneously in Canada by Random House of Canada Limited, Toronto.

A version of "Can We Tock" was originally published in *Mademoiselle.* All other essays, except "Gag Reflex" and "Voodoo Ergonomics," appeared in *Discover* Magazine

Grateful acknowledgment is made to Steve Martin for permission to reprint an excerpt from "King Tut." Copyright © 1978 by Plaid Men Music.

Library of Congress Catalog Card Number: 89-92611
ISBN: 0-345-36588-7

Text design by Debby Jay
Cover design by James R. Harris
Cover illustration by Dan Sneeburger

Manufactured in the United States of America

First Edition: May 1991

10 9 8 7 6 5 4 3 2 1

To my nuclear family, there through fission and fusion.

Acknowledgments

Thanks to the subjects and sources of these pieces, the women and men of the scientific community and its fringes. You were generous with time and expertise, and good sports all (especially considering that unpleasantness with the hamsters). Special thanks to the gang at *Discover:* in order of height, Marc Zabludoff, Paul Hoffman, Conrad Warre, Michael Witte, Clarence Reynolds, Bill Hively, Patricia Gadsby, Roseann Henry, Marcia Bell, and Carolyn Waldron. A laurel and a hearty handshake to Dennis Andersen, Magda Denes, Dalma Heyn, Stuart Krichevsky, Richard Marek, Andrew "Death to Flying Things" Partner, Betsy Rapoport, Lynn Rosen, the Writers Group—Mary Ellen Donovan, Lesley Dormen, Shelley Levitt, and Julia Lieblich—and Melanie Wyler.

Contents

□ ■ □

MINERAL

GENERAL

LIGHT ELEMENTS

Read This First

It'll Be On The Quiz

□ ■ □

I used to be the kind of person who didn't recall—or care—whether it's onto*logy* or onto*geny* that recapitulates Phil Donahue. I thought that gluons were false eyelashes and that Brownian motion was discovered by Betty Crocker (if I bothered to think at all about the random travels of microscopic particles suspended in a liquid or a gas). I was still a tad ticked off at Pope Gregory IX, who, in 1231, rescinded a ban on the study of physics. (Thanks for screwing up a terrific grade point average, Greg.) In other words, along with 95 percent of the American public (see page 9), I was a scientific illiterate.

Then I embarked on the explorations that became this book, partly because I felt uncomfortable living among natural and technological phenomena I found puzzling and unpredictable, and partly because someone paid me to do it.

Now, there are two ways to approach a subject that frightens you and makes you feel stupid: you can embrace it with humility and an open mind, or you can ridicule it mercilessly. I'm still deciding which I prefer.

In the meantime, I've pestered the experts for answers to the fundamental questions I guess we've all asked ourselves: Why does the sound of fingernails on a blackboard cause the willies? How can I stop my dentist from hurting me? Has anyone invented a really good way to twirl a squirrel? What do you call the little plastic thing on the end of a shoelace? What's all this stuff about a newly detected quasar at the edge of the universe? Do flatulent cows really contribute to the global warming trend? Where can I go to get mummified?

In the process I've been vouchsafed the joy of learning that the physical world is explicable, even effable; that the behavior of matter is governed by verifiable laws; and that scientists turn out to be pretty cool men and women once you walk a mile in their lab coats. I'm beginning to understand my place in the universe (you turn left just past the 7-Eleven). And I'm anxious to share.

No one can dodge death or taxonomy; the revelations in this book had to be arranged in a way that makes sense. How to present disciplines as diverse as physics, medicine, zoology, computer sciences, astronomy, and williology? Mendeleev's periodic table wouldn't make a good model; we're dealing here with only the lightest of elements. Something like Linnaeus's plant classification system? Too many hybrids, like psycho-acoustics and astrophysics. The *Victoria's Secret Catalog*? Not enough garter belts.

I decided to fall back on the simple organizational system that's served many of us since childhood: Animal, Vegetable, and Mineral. As a kid playing junior varsity Twenty Questions, I used to brood about items like cookies, which seemed to encompass all three. But rather than present a category called Cookieoid Hybrids, I've made it Animal, Vegetable, Mineral, and General. (Or maybe you'd prefer Potent Potables for $100.)

I've only scratched the surface in this volume; an infinity of questions remains. Among the ones I'm working on: How come they don't invent a digital clock-radio with a two-directional alarm-setting capability, so that if you suddenly get a crick in your finger and accidentally overshoot 7 A.M., you don't have to rock around the whole damn clock to get back there? Is time travel possible? Is that William Shatner's real hair?

I no longer believe life has to be a comedy to those who think and a tragedy to those who passed geology by writing a poem about lateral moraines. I haven't totally vanquished my ignorance; I've just stopped being so damned proud of it.

ANIMAL

□ ■ □ ■ □ ■ □ ■ □ ■ □ ■ □ ■ □

Is anything more interesting than our own behavior and that of our fellow creatures, from tiny one-celled organisms like amoebae to giant one-celled organisms like Kenny Rogers?

□ ■ □

If U Cn Rd Ths, U Undrstnd Scnce

□ ■ □

You say you don't know a proton from a crouton and you can't tell gravity from levity? You're not alone. A recent nationwide survey funded by the National Science Foundation shows that fewer than 6 percent of American adults can be called scientifically literate. The rest think DNA is a food additive, Chernobyl is a ski resort, and radioactive milk can be made safe by boiling.

"Only one in twenty adults knows enough about science to function effectively as a citizen and consumer when asked to help formulate public policy about issues like nuclear power or toxic waste," says Jon Miller, director of the Public Opinion Laboratory of Northern Illinois University, who conducted the study.

Certainly I have no right to be smug; I have, after all, been known to whine that if God wanted us to use the metric system, He would have given us ten fingers and ten toes. But, honestly, do you want these people voting on where to stash the leftover plutonium? More to the point, would you like to ride a chair lift or hoist a frosty, glow-in-the-dark malted with them? Well, they're us.

Miller's subjects, just plain folks, answered three sets of questions. The first measured their knowledge of the process of science. "What does it mean to study something scientifically?" people were asked; their responses had to include some mention of testing hypotheses, formulating theories, and using experiments or systematic comparative study. Those who answered correctly were then asked whether astrology was scientific. Only about 12 percent got both parts of the section right. (The rest must have been Geminis; they always have trouble with two-part questions.)

Twenty-eight percent passed the second part, which tested knowledge of scientific terms and concepts, mostly through true-false questions such as "The earliest human beings lived at the same time as the dinosaurs" (37 percent correctly answered false) and "Electrons are smaller than atoms" (43 percent correctly answered true). Also included were multiple-choice questions, like whether the earth travels around the sun or vice versa. Those who knew that the former is true were asked whether this event takes one day, one month or one year. 45 percent got it right; the rest had really enjoyed Mel Gibson in *The Day of Living Dangerously.*

Half the respondents aced the part that gauged their understanding of how science and technology affect their lives. That's where the radioactive milk came in; 36 percent weren't worried about its half-and-half life.

Age wasn't a factor in determining who was scientifically illiterate, although sex was; men scored somewhat higher than women did. Miller attributes the difference to the way science has traditionally been presented in schools—as a stereotypically male realm that girls are subtly discouraged from entering. The best predictor of a high score was having taken a college science course.

At least we're not getting dumber; Miller conducted similar studies in 1979 and 1985 and finds little difference in scores then and now. And a nearly identical survey of 2,000 Britons, conducted in 1988 by Oxford University, revealed that they know as little as we do. A slightly higher percentage of Brits grasp the impact of science on society, but twice the number of Americans—about half—know what software is. More important, we know that Benny Hill just isn't funny.

(We may soon have a chance to see how we measure up to the Japanese; Miller is trying to arrange a study there. You'll be relieved to know he thinks the Japanese scientific literacy rate will turn out to be "not much more than twice ours.")

Miller's figures are fascinating, but I wanted to learn more about what people thought they knew. So for a few months, wherever I traveled, I asked people questions about science—some from Miller's study and others inspired by recent news stories. I talked to people on airplanes, at restaurants, on the beach, and on the street—basically, anyone who didn't pull a knife when I said, "Hi, what's the ozone layer?" My study of nearly 150 unarmed Americans, ranging in age from three and a half to sixty-seven, was rather informal but highly revealing. I was heartened to see how many respondents (just under oodles) could define the scientific method but demoralized to discover how few seem to apply it to their lives. All answers are totally serious, except those followed by the words "Just kidding."

First came my tricky reworking of one of Miller's questions: "How," I asked, "did the earliest human beings fight off dinosaurs?" You and I know that the last of these creatures vacated the planet more than fifty million years before *Fredus flinstoniensis* moved in. But only about half the folks I polled had gotten the news. A twenty-eight-year-old actress/waitress at a Manhattan restaurant insisted that coexistence wasn't a problem since dinosaurs were vegetarians. According to an eighteen-year-old advertising major at San Francisco State University, "Early human beings would hide in caves, set up traps, or play dead when dinosaurs came upon them." (If a dinosaur came upon me, I'd probably play dead, too. Talk about your sticky situations.)

Thank heavens for Andrew, a three-and-a-half-year-old San Francisco nursery school student majoring in fire engines, who declared, rather condescendingly, "There *weren't* any people. I saw a show." His father explained that they'd recently seen a play about dinosaurs at the University of California's Lawrence Hall of Science in Berkeley. Parents take note; otherwise your kid could end up like Buddy, an Alabama man visiting Pensacola, Florida, the southern tip of what the locals refer to as the Redneck Riviera. "First," Buddy said, "early man called the dinosaurs names—'you ugly bastards!' Then he ate their eggs." He took a pull at his Bud Light and inhaled a handful of Zapp's Crawtater Chips. "Just kidding."

More people correctly defined the ozone layer than any other scientific term; a librarian at the University of Virginia and a retired auto mechanic in Portland, Oregon, even knew that it's oxygen with an extra O. But a twelve-year-old member of a Kilgore, Texas, church group identified it cryptically as "the middle layer." I believe she confused *ozone* and *Oreo*, a slip even Nobel laureates sometimes make.

When asked to identify Isaac Newton, nearly all adults surveyed mentioned gravity (without going into further detail), and they correctly linked Charles Darwin with the theory of evolution. But a St. Louis man told me, sans elaboration, that Darwin was "known for his use of force," and a twelve-year-old Kilgore boy stated confidently that the two men were actors—scientific method actors, no doubt. I know I especially enjoyed Chuck "the Enforcer" Darwin in *The Originator.*

Recognizing Chernobyl would be a lead-pipe cinch, I thought, but I included the item anyway since in a recent Gallup poll a fair number of high school students thought it was Cher's real name. More kids drew a blank on this one than

I expected. And a thirty-nine-year-old accountant returning to San Francisco State as a freshman really did think it was a ski resort. My favorite answer, from his megamellow nineteen-year-old classmate: "A Russian nuclear power plant that totally blew up, dude."

DNA, you'll be interested to learn, is a sickness (a nine-year-old girl from Little Rock said that), the smallest known molecule (according to a twenty-six-year-old woman from Philadelphia), a food additive (another Philadelphian, recently overheard in a Chinese restaurant saying, "I'll have the Hunan beef, but go easy on the DNA"), and an airport. (This puzzling answer, from a Kilgore third grader, was explained by an older friend. "I think she's thinking of DFW—Dallas–Fort Worth.") Among the cognoscenti, the term *genetic blueprint* won the prize for Most Bandied About. An architect from Baton Rouge made a fashion statement: "It's the double helix that holds all our genes together."

Oddly, this question yielded the only vaguely hostile answer, from the head of a New York public relations firm who hollered in a French restaurant, waving his *boudin* for emphasis, "I don't know what DNA is and I don't want to know!" Most people were either apologetic or oddly proud of their ignorance, but always good-natured.

The hardest question on the survey, one I myself couldn't answer extempore even if you threatened to make me watch an endless loop of "Hello, Larry" reruns, was "Explain how television works." You can guess how many wags from coast to coast felt compelled to say, "You push the button"; I got quiplash. The typical response was in the ballpark, but not quite complete, like this one from a thirty-year-old woman who's art director of a Baton Rouge business publication: "Radio waves are sent through the air and electronically reorganized in your television." A Texas eight-year-old said simply, "Microchips," which was more than one child's answer to more than one question. "It's done with mirrors and broadcasting," said a San Francisco man. And on the Redneck Riviera, one answer sparked a lively debate. "Radio waves are sent and received," said good old Buddy. "But if they're *radio,*" asked a companion, "how do they know when to turn into a picture?" "How," said Buddy sagely, "does a thermos know when to keep something hot and when to keep it cold?"

I especially enjoyed the responses of Gail Sipes's third-grade class at the A. B. McDonald Elementary School in Moscow, Idaho, a generally canny crew that includes one young *civis indignatus* who, when asked which is heavier, a pound of bricks or a pound of feathers, wrote, "Bricks!! Of Course!!!" The kids also revealed that software is either "plastic knives and forks" or "clothes that keep us warm," and that the ozone layer "separates bad air from good air."

How alarmed should we be at these answers? Does an inability to explain DNA make you an awful person? Is it more important to recognize the name Chernobyl or to have the psychological savvy to judge the guy with his finger on the nuclear button? Just what, in this large loaf of data-nut bread that is the modern world, do we need to know?

Miller worries that citizens who think the answer to the question "How long does it take the earth to revolve around the sun?" is "A long time" will have trouble making public policy decisions involving science. (Political leaders could do better, too, he believes. "I think members of congressional committees who deal with science would do quite well on my test. But our last president would have failed the first question on my survey; he didn't know that astrology isn't a science. That alone qualifies him as scientifically illiterate and should have disqualified him from being president.")

Look, we can't all be Einstein (because we don't all play the violin). At the very least, we need a sort of street-smart science: the ability to recognize evidence, gather it, assess it, and act on it. As voters, we're de facto scientific advisors. In the next few years we're going to be making, directly or indirectly, vital decisions about the greenhouse effect, acid rain, the pesticides that taint our foods, genetically engineered organisms, how much to spend on mending our torn ozone layer. (Or Oreo layer.) If we don't get it right, things could go very wrong. The oceans will rise, the trees shrivel, the snow turn to steam; nothing will taste good. If we don't get it right, they'll be shedding their thermal software in Antarctica.

Herd It Through the Bovine

□ ■ □

Cattle belch mightily every minute and a half.

Go ahead. Make droll remarks: "Gee, they don't even say excuse me. Where were they raised, in a barn?" Slap your knee and wheeze with laughter. Enjoy yourself while you can. The planet is about to be cowed into submission.

These burps dispense methane gas; up to 400 liters of methane a day erupt from each of the world's more than 1.2 billion head of cattle (a figure that the International Farm and Agricultural Organization no doubt arrived at after counting hooves and dividing by four). "So," you gasp between guffaws, "what?"

Methane happens to be the second largest contributor to the greenhouse effect, the worldwide warming trend caused by an increase in the earth's atmosphere of gases that admit sunlight but trap heat, like the windows of a *greenhouse*—hence the name "worldwide warming trend." Carbon dioxide is the big culprit. In the last 100 years, atmospheric CO_2 has increased by about 15 percent—more than in the previous 200 millennia—because of the burning of fossil fuel and rampant deforestation; we've destroyed the plants that would have converted thousands of tons of CO_2 into our friend oxygen. According to chemists at the National Center for Atmospheric

Research in Boulder, Colorado, CO_2 deserves 60 percent of the blame for the greenhouse effect. The rest of the responsibility lies with trace gases, including methane. And most of that methane is produced by cows.

Cows, in fact, are essentially large fermentation vats in leather. Conveniently for humankind, they and other ruminants—four-stomached cud-chewers—are able to turn unsavory foodstuffs like grass and hay into meat and milk, thanks to bacteria and other microbes that inhabit the rumen, the largest of their four stomach chambers. Methane is a byproduct of fermenting fodder. Luckily for farmers and ranchers, most of the gas escapes in burps rather than from the business end of a loaded animal, though enough is found in manure to put a kick in the dairy air.

You still don't get it, do you, you big four-stomached cud-chewer? While you're tittering about emission control devices for yaks, life as we know it may be coming to an end. Mrs. O'Leary's lantern-kicking cow charbroiled a mere metropolis. Today's belching bovines may heat up the whole world. A rise in global temperature of a mere three to nine degrees could, some scientists think, melt the polar icecaps, thereby swelling our seas, drowning coastal cities, dramatically changing worldwide weather patterns, screwing up ocean currents, killing a major chunk of the world's fish population, raising the prime interest rates, and making your face break out.

Luckily, people like biologist Donald Johnson of Colorado State University are working to make the world safe from bovine eructation.

That's not what he set out to do. "Basically, I was trying to find ways to produce the same amount of cow using less feed. Since about 6 percent of what cattle eat is wasted in methane burps, my main thrust was to decrease methane losses so I could increase the efficiency of cattle growth."

Johnson experimented with drugs called ionophores, growth

enhancers chemically similar to antibiotics. "Ionophores change the way cows ferment feed in the rumen by inhibiting certain kinds of microbes," he explains. "In general, the ionophores worked well. We got the same growth with 6 to 7 percent less feed, and methane loss was decreased by 4 to 25 percent."

How the heck can he be sure? I hear you cry. Johnson collects data from volunteer steers ensconced in large respiration chambers. "These are basically big plastic boxes with air flowing through to ensure that the animal is comfortable. We measure methane entry and exit from the chamber."

The larger implications of his work hit him only recently, Johnson says. "At an EPA workshop on trace gases, I discovered that atmospheric scientists consider bovine methane emission to be a major contributor to global warming. And since the cattle population is increasing faster than the human population, methane may become an ever-growing problem. That's when I began to think that while improving efficiency of production, we might also be able to decrease global warming."

Some of you aren't taking this seriously. Perhaps you've heard that Johnson occasionally amuses his freshman classes by collecting gases from the rumen in a bag and lighting the fumes. Okay. Mutter (and fodder) your feeble cracks about udder nonsense and bunch of bull and bum steer. Then listen to atmospheric chemist Patrick Zimmerman, who heads the National Center for Atmospheric Research's biosphere-atmosphere interactions project: "Methane is an important greenhouse gas, even though it's found in concentrations of only 1.7 parts per million, compared with 365 parts per million for CO_2. But methane is transparent to certain wavelengths of light that CO_2 isn't; small amounts can cause a big effect. And methane emissions may be tougher to control than CO_2 emissions, which are tied to energy production and can be conserved. Methane production is, for the most part, natural, linked to the food supply.

"We know that methane is increasing at about one percent a year, and right now we're identifying the sources. You produce a little bit when you burn a jungle; there's a little bit in car exhaust and leakage from natural gas wells. But over 80 percent is biological and natural. Any time you have microbes that can only live in the absence of oxygen, methane is produced—in swamps, rice paddies, landfills, or the guts of animals."

Not surprisingly, India is the world leader in bovine methane production. Sheep belch even more than cows, but there aren't as many of them. Camels are not blameless, but it would be hard to fault giraffes, the ruminant with the tallest emission tower. Then there's this termite thing.

"Only recently have we realized that termites have the potential to affect the earth's climate," Zimmerman says. "Termites eat material that no one else wants, and microorganisms in their gut efficiently break down carbon and produce methane. The amount of carbon recycled is huge—as much as two-thirds of the carbon on land ends up going through termite gut, and one percent of that becomes methane. We haven't looked at how the gas is released; probably it diffuses out through the spiracles, the breathing holes in the insects' sides.

"A termite queen can lay 60,000 to 80,000 eggs a day. No one knows for sure, but we think there are three-fourths of a ton of termites per person on earth." Stand by for instructions on where to pick up yours. "And deforestation is increasing their numbers," Zimmerman adds. Live trees produce substances that ward off termites, but cleared areas offer an ideal habitat—piles of defenseless chips to snack on. In an ex-forest, he says, the termite population grows by a factor of ten.

But which are worse, termites or cows? "It depends on

whom you ask," says Zimmerman. "I think termites are a larger source of methane than cattle, but many of my colleagues who are reasonable, rational people would say that termites are a lesser source. We don't have good enough data. Right now we're trying to solve the source-strength controversy by discovering the unique isotopic 'fingerprints' of methane from various sources—rice paddies, cattle, termites—then trying to find those fingerprints in atmospheric methane so we can tell which sources contributed." Until he and his colleagues know more about the sources of methane, he says, they can't predict where the global warming trend may lead. "Change is part of geochemistry. But the climate changes associated with the greenhouse effect are happening faster than changes associated with ice ages. That's the scary thing." Zimmerman thinks Johnson's work with ionophores appears to be a promising step in the fight to keep Club Med out of Siberia.

Johnson is now exploring the persistence of ionophore action. Just how long does it inhibit methane production? "Our studies show that after about sixteen days in an animal and fifty days in a laboratory system, microbes adapt to the presence of the drugs," Johnson says. "Methane production returns to near normal, though we're still able to use less feed. We're on the right track, but further experimentation is required. Recently it's been shown that fungal organisms and protozoa in the gut are involved in methanogenesis. Perhaps if we can inhibit them, we can inhibit methane."

Johnson thinks scientists shouldn't pooh-pooh the idea of working with manure, and not just burps (although he hastily pleads prior commitments himself). "Curiously, chicken droppings have been totally ignored. I think we should be investigating the amount of methane in manure of all livestock. I'm quite sure there are significant amounts in many systems."

Though Johnson is serious about the dung approach, he still believes the most elegant solution to the methane problem would be high-tech, not high-dreck: "Ultimately, if one could genetically engineer a gut microbe to do its job with minimum or no methane production, that would be ideal."

Should we be trying to create a new kind of cow, a bossy nova, as it were? Animal physiologist Harry Colvin of the University of California at Davis, who's been studying eructation pathways since 1953, thinks we ought to be happy with the fat and gassy animal Nature has given us. "The whole evolutionary reason for cattle is to use very poor-quality substances like hay through the process of fermentation. I don't know how to alter methane production without altering the nature and purpose of ruminants."

Sure, many questions remain unanswered. We need to determine the relative contributions of humankind, kinekind, and termites to the change in our weather. We need to know more about the long-term effects of ionophores on bovine methane production. But if I've made you understand that cows aren't just hip little decorations for designer ice cream boxes; if you've learned to have a healthy respect for them as the grass-chewing, gas-spewing menaces they really are; if you've accepted *spiracle* as a cocktail party buzzword, well, just maybe we'll all be a little better prepared for *La Belch Époque.*

Gorillas Just Want to Have Fun

□ ■ □

"Good gorilla sperm is so hard to find!" sighs Francine "Penny" Patterson, Ph.D., and I guess there's not a woman in America who'd disagree with her. She apologizes for interrupting dinner at a restaurant south of San Francisco to answer a telephone page; a *Washington Post* reporter has managed to track her down on one of her rare nights away from the secluded Woodside, California, farm where she and her companion, Ron Cohn, a molecular biologist, live with the world's only two gorilla conversationalists, Koko and Michael.

The reporter wanted to know how things were going in the quest to get some of that recherché semen to eighteen-year-old Koko, who for some time now has been telling Patterson she wants a baby. Koko communicates in American Sign Language, which Patterson began teaching her in 1972, when Koko was still living at the San Francisco Zoo, where she was born. The two have been together ever since. What began as Patterson's thesis project for a doctorate in psychology has become Project Koko, a pioneering investigation into the ability of gorillas to acquire language. Koko's vocabulary is equivalent to that of a five- or six-year-old deaf child. Michael, who joined the family, and the conversation, eleven years ago, knows about half as many words.

"When Koko asks about a baby," Patterson says, "I tell her, 'you'll have to get Michael to help you with that.' " But Michael, fourteen, is being something of a "devil stink rotten toilet"—one of Koko's favorite epithets—about it.

Actually, that's being unfair to the handsome, 400-pound silverback; he's as ready to commit as any guy. The roadblocks to romance aren't his fault. For one thing, the environment isn't conducive to the making of whoopee. The two live in fair-sized, side-by-side trailers with outdoor enclosures, but gorillas need space to get into the mood. In fact, Patterson and Cohn would like to move the gorillas to a larger preserve in a warmer place. Patterson recently spent six days—the longest she's been away from Koko during the past fifteen years—in Tahiti, where she discussed with Marlon Brando the use of one of his coral atolls. The move, like all Project Koko activities, would be funded by private donations from the 25,000 members of the Gorilla Foundation.

Another problem is that for Koko, Michael may be ape priori off-limits. Orphaned in Africa by hunters, he joined the family when Koko was five and he was three and a half, early enough for her to regard him as a pesky younger brother. Wild gorillas live in family groups, and incest is apparently unacceptable.

At times, in the last few years, the pair has seemed about to transcend trailer and taboo. Koko has invited Michael, in sign language, to "walk up her bottom," a phrase that is the gorilla equivalent of "New in town, sailor?"

Says Patterson, "At one point there were some very clumsy attempts at mating, but they were interrupted when Koko was attracted to a human male who was working on the project. If she were in the wild, that's how she would bond to a male: she'd see him in the distance and she would leave her family for him. It's just that there aren't gorillas out there, but humans."

Koko is having trouble understanding that human men are

unobtainable, and it's frustrating for her, Patterson says. "We've explained that it's not going to work." She's even shown Koko videos of gorillas in sexual situations, but *Debbie Does Rwanda* hasn't helped yet.

" 'Michael's the right one for you,' we tell her; and she has actually taken it to heart and approached him. He's interested, but his instincts are telling him that he must dominate her. And her instincts are telling her that though he outweighs her by a couple of hundred pounds, he's the kid brother and she's not going to be dominated."

Patterson isn't sure how much time Koko has left. "I've heard a report that there's a thirty-one-year-old cycling female [that's menstrual, not Schwinn] in a zoo, but we don't know if she's able to conceive. Koko may have longer than we thought. Science is still learning about gorillas; we don't even know if they have menopause."

What's a gorilla to do? Run a personals ad? ("Swinging single female, attractive, successful, likes 'Sesame Street,' peanut butter, long walks, long arms; seeks sensitive, caring, vegetarian knucklewalker for serious relationship. Opposable thumbs a must.")

Sure, it's easy to make cheap jokes at Koko's expense. But the matter of offspring is as serious for science as it is for the wistful madonna-wanna-be. Koko's child would, besides helping to replenish the world's dwindling gorilla population, allow Patterson to study a second generation of gorilla communication and see whether Koko passes on her signing skills. Patterson reports that Koko already molds the hands of baby dolls, and Koko and Michael talk to each other in sign language when they're together in private, though they stop when they see that they're being watched.

Artificial insemination is the logical answer to Koko's woes. But obtaining semen from Michael would require general anesthesia (would *you* like to try another way?), a difficult procedure involving health risks that Patterson doesn't want to take. Even if Michael's sperm were easy to get, it might not do the job. Captivity, however cushy, seems to create hormonal imbalances that may render some male gorillas sterile. Patterson has spent months tracking down a donor who's logistically and hormonally suitable.

Colossus looked, at first, as if he were the answer to a maiden's prayers. Recently the twenty-one-year-old male gorilla was moved from a small New Hampshire animal park to The Zoo in Gulf Breeze, Florida. Since he was going to be anesthetized anyway for a complete physical, zoo officials agreed to collect some semen for Koko (using an electronic ejaculator, if you must know). But the ape's sperm proved to be of poor quality. Enter the top-seeded Ivans, star attraction at a Tacoma, Washington, shopping mall. Patterson hopes to inseminate Koko with the twenty-four-year-old gorilla's sperm.

Having a child by Ivans may boost Koko's chances of having another, by Michael, Patterson thinks. "We're hoping that there might be chemistry if we add another gorilla—a baby, or perhaps another female, because a male in the wild has a harem, a group of females. A pair isn't a functional social unit in the wild."

At forty-one, Patterson's a tiny, lovely woman who looks like Meryl Streep playing Penny Patterson in *Koko: The Movie.* Two documentaries have been made about Koko, but no feature film; it seems to be a natural. Patterson laughs at the idea. "Who would play Koko?" she says. Although Streep could, of course, handle both parts, Koko herself seems an obvious choice. "There's no way she'd take direction," Patterson says. "Koko's not trained. She does what she wants. Her external controls are what we think of her and how we respond to her. Punishment comes in the form of verbal threats—'you won't get your sandwich if you're bad'—or time-outs—leaving her alone for ten minutes."

On the spring day I visit Woodside, what Koko wants is satisfaction. She's in heat and behaving the way any horny young individual might; she's been experiencing UMS (ugly mood swings) and engaging in bottom bouncing, a form of primate display activity—a combination attention-getter and masturbation—that looks the way it sounds. Mama said there'd be days like this.

The girl's got it bad. Recently, Patterson reports, a deaf human baby visited the compound. "Koko asked me to bring the baby into her room. She wanted to hold that baby so much!" This morning, Koko and Mary Kennedy, one of her teachers, were showing her baby dolls how to sign. "Koko purred when I praised the babies, and she pretended to nurse them," Kennedy says.

In urgently communicating her babylust, Koko once again forces *Homo sapiens,* the only animal that shops at K mart, to rethink what it means to be human. Koko has used language to express grief; she still signs "cry frown sad" when she sees pictures of her kitten All-Ball, who died in 1984. She expresses affection and compassion—she's especially tender with newborn and elderly humans—and remembers her dreams. She's a poet, able to sign words that rhyme in English, like *All-Ball,* a name she created, or *bread/red/head,* and a maker of metaphors: She calls her Pinocchio doll an "elephant baby," a mask is an "eye hat," a cigarette lighter a "bottle match." She can write her name, a skill she picked up on her own, and she likes to be read to—scary tales and stories about cats or gorillas, especially her. She's also learning to read on her own.

"What gave Penny the clue that Koko could read," Cohn recalls, "was that she could distinguish objects that were identical except for the words on them. Now she works with flash cards, games, and alphabet pasta." Apple is working on a computer that would allow Koko to pound out written words.

Koko laughs a sort of voiceless human guffaw at her own jokes and others'. "Her humor is primitive; she finds incongruity funny, the way a young child might," Patterson says. "I once asked her what was funny and she put a toy key on her head and said it was a hat, and pointed to a puppet's nose and said it was a mouth." Koko also cries, a sort of heartrending wooo-wooo, when she's sad or lonesome. And she lies, once blaming a broken sink on a human volunteer. She's even thought about where gorillas go when they die: "Comfortable hole bye."

Koko and Michael each have strong, unique personalities. Koko is more outgoing, with a sense of her own celebrity. Michael's dreamier, the shy, musical type. Both prefer classical music to rock. "Michael's creative with rhythms," Patterson says. "He can recreate rhythmic patterns on different parts of his body." He's especially attracted to the children's songs on "Sesame Street" (though a turtle puppet on the show once terrified the big guy). He's drawn an excellent picture of a spider.

Patterson sees no sharp defining line between animals and humans. "Their characteristics and ours merge. There's no one thing I can point to and say, 'There, that's the difference.' It's a question of degree."

A few critics, she notes, insist that Koko is merely imitating her human teachers. "But if Koko starts conversations, then she can't be parroting—and she does, close to 50 percent of the time. If she makes up her own new words in sign, where are they coming from if she's not thinking, if she's just imitating? She does that routinely. She makes up gestures to convey new concepts that she's trying to express.

"The recent ones she's been doing have been based on sounds of English words. For instance, she's invented a sign for *bring,* which is based on the sound and sign for *earring.* She understands English, so the way we can disambiguate some of these things is to simply say to her, 'Tell me the word for—' and ask her to fill in the blank."

Patterson says we can visit Koko in spite of her spring fever. We walk from the trailer that serves as headquarters for the Gorilla Foundation, past the gently decaying farmhouse where Patterson and Cohn live, to Koko's trailer.

I'd asked Patterson what kind of presents Koko might like. She'd already requested cigarettes and a gold tooth for her birthday, July 4. "Oh, a scarf, earrings, lipstick, hair combs," Patterson had said. "She likes lady things."

Koko's in her enclosure, the model of decorum. Her face is beautiful, regally wrinkled but mischievous, too. She could be a totem animal worthy of worship or the world's wisest toy. I sit on a bench a few feet away.

Koko, who'd be about 5′5″ if she stood up straight and weighs around 250, points to her elbow, a sign she's created that Patterson thinks means "come closer." (Patterson says it can take as long as a year for her to decipher some of Koko's verbal inventions.) I crouch near the mesh of her enclosure. Koko once said, in response to a question from chimp expert Jane Goodall, that she prefers humans who approach her to be lower than she is. We check each other out for a while. Koko points to her teeth. Patterson interprets her signs: "She wants to know if you have any gold teeth." The gorilla gazes with frank interest at a pair of crowns that recently bought my dentist a summer home. She signs again. "You have red barrette there." I'm wearing a plastic hair clip in her favorite color.

"Oh," I say without signing, "as a matter of fact, I brought that for you." Koko holds out her hand and I give her the clip. She has trouble fastening it to her head. "Come here, honey," Patterson says, and signs "I'll put it on for you." Koko brings her large head up next to Patterson, who snaps the clip in place.

Then Koko spots the white plastic bag I brought, sitting on the bench. "What brought you there?" she signs.

First I hand her a pair of ultracool, aerodynamic wraparound glasses, also in red, with lenses that make rainbows when you look through them. "Dark glasses," she signs. "What a smart girl you are!" Patterson says in delight. Koko blows on the lenses, wipes them against her arm fur, and puts on the glasses. "They make colors," Patterson signs. She tells Koko how pretty she looks. With her natural simian flattop, Koko resembles Brigitte Nielsen or Grace Jones. She makes a circuit of the enclosure, very pleased with herself. "More," she signs, and I hand over a small, pink stuffed rabbit.

Up to now, Koko's expression has been that of a blasé teenager browsing at a mall. Suddenly she smiles, her eyebrows raised. "Baby!" she signs. She hugs and kisses the stuffed animal, signing "Baby Koko love." After a brief interlude of rapture, she places her baby on the ground, points to my shopping bag, and signs "Barrette there." The canny creature had heard me tell Patterson I had two more clips for Koko. I hand her a white one with a flower lightly embossed on it. She signs "Flower." Penny tells her she's very observant.

Koko, wearing red, white, and blue hair clips, races around her enclosure and bangs on the walls. "No," Patterson says sternly. "That's not allowed. We need to say good-bye now. You know about banging." Chastened, Koko calms down, and we stay. She wants the empty bag; that, too, ends up on her head. She sniffs my strawberry hand lotion, kisses my knuckles, and, pointing to a capacious armpit, signs "Tickle me there." In fact, she invites me to join her in the enclosure for a game of tickle-chase.

I've been a Koko fan for years; I used to visit the San Francisco Zoo when she was a tiny apelette in baby sweaters.

Being invited to play by Koko is like being asked to dance by Fred Astaire.

But it is not to be. "Honey," Patterson tells her, "we're not going to chase because there was a bang." A dutiful Koko tidies her yard, bringing Patterson the bunny baby and hair clips; Patterson praises her for being good and polite. "Okay, little one," she says. "I'm going to let you have your privacy. We're going to say good-bye now."

I still have questions I'd like to ask Koko. Patterson says she has better luck "interviewing" Koko over the course of a few days, so I leave her a list of questions. It comes back, eventually, with Koko's answers and a sample of her "handwriting" in purple ink.

Question: *What frightens you?*
Koko: Bite.
What do you dream about?
Lip [her sign for females, from *lipstick*] pick eyes.
What is a nightmare?
Devil look.
What do you remember about when you were a baby?
Thirsty good . . . stomach.
What's the hardest thing to learn?
Hard Koko good quiet.
Would you rather be with people or gorillas?
People good.
What kind of presents do you like to get?
Fake tooth [gold crown].
Can you think of anything else you want?
Bite corn.
Who are your good friends?
Devil [Koko's nickname for Ron Cohn].
Would you like to have a baby?
Koko love good. Baby good.
Why do you want a baby?
Nipple mouth.

The day after my visit, the archive where every communication is logged shows that Koko told Cohn this story about the event: " 'Hair comb head Koko. Comb hair comb.' [She signed *comb* many times, all over her head.]" Okay, she ain't Jane Austen, but I think it has a compelling narrative thrust.

And nearly three years after my visit Patterson and Cohn still aren't sure when or if they'll become grandparents. ("We function as a family," says Cohn. "A weird family, but a family. She's identified us as her mother and father." Presented with photos of her biological mother and father, an unrelated gorilla pair, and Patterson and Cohn, Koko picks the humans.) Koko seems ready for motherhood. "She's becoming much more responsible," Patterson says, "cleaning her room without being asked, handing me magazines she knows she's not supposed to have overnight." She continues to mother her cat, Smokey, the now-grown replacement for the lost All-Ball. Perhaps Ivans will come through with the goods after all. But for now, Koko remains just another career gorilla facing the ultimate problem of young upwardly mobile primates.

Personal Beast

□ ■ □

For the past several millennia, dogs have pretty much had the Man's Best Friend market cornered. Lately, however, thanks to a sort of demographic Darwinism, several strong contenders for the title are nipping at the heels of the chosen species.

The dog emerged as protopet in Mesopotamia, where our nomadic ancestors first began living in villages about 12,000 to 14,000 years ago. (And if you think it's hard to paper-train a puppy, imagine having to use stone tablets.)

"Domestication was an urban event," explains Alan Beck, Ph.D., director of the University of Pennsylvania's Center for the Interaction of Animals and Society. "The garbage generated by these new high-density human communities probably attracted wolf packs; villagers may have bred the pups into pets with the idea of making peace with the pack. These domesticated creatures could bark a warning when nondomestic animals came around, and also help with trash control. And perhaps one of the earliest reasons for breeding pets was for companionship; the desire to nurture is part of human culture." (Those days, when Hector was a pup, probably marked the first time in history that a human being uttered the words, "Aw, Ma, can't I keep it?")

Now a new kind of urbanization is creating a new kind of pet. People are choosing animals that better fit a busier life and a smaller dwelling. Is there room for Fido in Mondo Condo, a world of two-income families with less time but more discretionary income? Folks want pets that are small and independent, pets that offer them a way to announce their individuality in a crowded, standardized world. Miniaturization, convenience, chic—the pet of the nineties has many of the same fine qualities as an under-the-counter microwave or a car phone. Here, hot off the Ark Nouveau, are the exotic animals that busy Americans simply must have.

Some of you may be seeking a companion that's low-maintenance, affectionate, cute (though swaybacked and paunchy), and perfectly content curled up in front of the TV, pigging out on junk food. (Yeah, yeah, I know—you're already married to it. You're a scream, honestly.) I mean the Vietnamese pot-bellied pig, the nation's top-selling Asian import. A black, beagle-sized porker that's also called a Chinese house pig, it looks like a cross between a hog, a honey bear, and a hand puppet.

"If you can keep a poodle, you can keep one of these pigs," says Fredericka Wagner, co-owner, with husband Bob, of Flying W Farms in Piketon, Ohio, which sends more of the little piggies to market than anyone in the country. "They're very appealing. Full grown, they weigh up to 45 or 50 pounds and stand about 12 to 13 inches high. They graze instead of root. Ordinary pigs have a long, straight snout; this one has a short, pushed-in, wrinkly nose. Its little ears stick straight up like a bat's and it has a straight tail that it wags like a dog's. It barks, too, and it can learn tricks—to come when it's called, sit up and beg, or roll over and play dead hog instead of dead dog." And it's easier to housebreak than a cat, Wagner says. Her three little pigs, Choo-Choo, Matilda, and Hamlet, like to sit around with the family and watch movies on TV (presumably *Porky's I* through *III*), and they enjoy supplementing their diet of Purina Pig Chow with candy, brownies, peanuts, and potato chips.

When the pigs arrived two years ago from Vietnam (by way of Sweden and Canada, after three years of red tape), the Wagners already ran a midget menagerie, selling impossibly cute 18-inch miniature sheep (perfect for baby sweaters, easy to count for insomniacs), pygmy goats, miniature donkeys, and championship miniature Arabian horses. (Apparently there are no bonsai bovines, or the Wagners would have them.) "Miniature horses are bred down from larger horses—anything from huge Belgian draft horses to Shetland ponies—a process that can take a century," Wagner says. "In our experience, to reduce a horse from 48 inches to 34 inches takes six generations—about twenty years. You can do it in only three generations if you have a 30-inch stallion, but you'd have to dig a hole to put the mare in."

Wagner first heard about the pigs from a friend in California; reportedly they'd been imported by Vietnam veterans who recalled the friendly critters from their tours of duty. Though Wagner was instantly attracted to them, she wasn't sure she'd get the business off the ground. But the swine flew. "We can't keep up with the demand," she says. "In the first eighteen months we sold a hundred pigs." Wagner is boarish on pot-bellied pigs as an investment. "Not even the stock market will pay you back as fast as these pigs will. I've had several retired people buy them to supplement their income. They breed at six months [the pigs, not the retirees] and it takes three weeks, three months, and three days for them to have babies. By the time your gilt is a year old—a gilt is a pregnant sow—she's given you her first litter of pigs and you have your investment back three times over." Since you can only have gilt by association, those who want to breed pigs must buy an unrelated pair for $2,500. "We expect that to go up to $3,000, because the demand is so great," says Wagner, in hog heaven. "We've sold them to everyone from poets to princes—we shipped some overseas to a Saudi Arabian prince. Stephanie Zimbalist, the actress, has one."

Stephanie Zimbalist! A recommendation, indeed. But what about having the same pet as *Michael Jackson*? For part of his personal zoo (boa, deer, chimp/valet, glove, and, nearly, the Elephant Man), Jackson has chosen a taller order of hip creature, the llama. His is one of about 15,000 in the country. "But in years to come, we'll see more and more of them in the average home," says Florence Dicks, owner of the Llonesome Llama Ranch in Sumner, Washington. "They're great hiking companions, their wool is increasingly in demand, and they make wonderful pets. They've been part of domestic life in South America for centuries. I think of them as one of life's necessities."

Dicks, who runs Llama Lluvs Unlltd., the world's only Llamagram delivery service, notes that the recent llifting of a government ban on imports will increase the llama population; most American-born llamas are descended from a single herd owned by William Randolph Hearst.

"They're very gentle," Dicks says. "Many of my fifteen llamas lived in the house for their first year. They're easier to house-train than a cat." (You know how all weird meat is described as "sort of like chicken"? Apparently all weird pets are easier to housebreak than a cat.) Dicks explains, in more detail than necessary, that llamas are what's called communal voiders—a great name for a rock band. Spread some llama droppings where you want them to go, and, in the comradely way of communal voiders everywhere, they will use that spot forever after. "We've shared our bathroom with llamas for five years, and they've never had an accident," Dicks says. Okay, they squeeze the toothpaste from the top, but nobody's perfect.

"I train my llamas to hum—that's the noise they make—

when they want to go outside," Dicks reports. "They communicate by tone variance. If they're relaxed, there's a musical quality to the hum. If they're stressed, you can hear the anxiety. I have a llama who hums with a rising inflection when he's curious."

Full-grown llamas can stand over six feet tall and weigh up to 500 pounds. A male costs between $1,500 and $15,000. (And at a recent auction, a male said to have outstanding stud qualities—he always sends a thank-you note—fetched $100,000). A gelded male will set you back $700 or $800, a female about $8,000. The only bad thing about llamas, Dicks says, is that you have to clip their toenails every three or four months. Which doesn't sound like a big deal if you're already sharing a bathroom.

But more and more of us have life-styles and living spaces—to use a pair of expressions even more nauseating than the word *pus*—that can't accommodate a dog, let alone a llama. I know I barely have room in my apartment for a pet peeve. Hence the proliferation among city dwellers of ferrets, dwarf rabbits, and birds. According to the American Veterinary Medical Association, birds are the fastest growing pet category. Their numbers increased 24 percent between 1983 and 1987, from 10.3 million to 12.8 million. Talking birds, like Amazon parrots, are especially sought after, reports veterinarian Katherine Quesenberry, head of the exotic pet service of the Animal Medical Center of New York. (And I guess if you crossed these South American birds with llamas, you'd get Fernando Llamas, six-foot communal voiders that squawk, "You look *mah*velous." Cheep gag.)

The ferret, a more personable cousin of the weasel, has been bred in captivity for a century, mostly for lab research. But its popularity as a pet has steadily risen over the last decade. Says Tina Ellenbogen, a Seattle veterinarian and information services director for the Delta Society, a national organization dedicated to the study of human-animal bonding, "People become attached to ferrets because they have a lot of personality. They're small, clean, and amusing." (It's sometimes hard to tell when people are talking about ferrets and when they're talking about Dudley Moore.) You can walk them on a harness or let them play on a ferrets wheel. They're easily litter-trained, says Ellenbogen—easier than a cat, I imagine—and statistically less likely to bite than a dog is. Males are unpleasant if you don't remove their stink glands, and females are sexually insatiable until you have them fixed, but hey.

Maybe you're a person who doesn't understand all the sound and furry over mammals. Maybe you'd rather see something in cold blood.

The reptile of the hour is the African Old World chameleon. "Having one is like owning a dinosaur!" says Gary Bagnall, head of California Zoological Supplies, one of the five largest reptile distributors in the country. "They look truly prehistoric. Their eyes move independently and they have 10-inch tongues with stickum at the end for catching insects. The base color is green, but they can blend into their surroundings by changing to yellow, orange, white, black, brown, and sometimes blue." The 6- to 10-inch chameleons start at $35; a foot-long variety, called Miller's chameleon, goes for $1,000. "We get only about four of them a year," Bagnall says. "There's a waiting list."

The nation's most sought-after amphibian, according to Bagnall, is the poison arrow frog, a tiny (less than an inch long), jewel-like native of South America that comes in orange and black, yellow and black, or blue. The really great thing about the poison arrow frog is that if you boil up about fifty of them, you get enough of the toxin they secrete to brew a dandy blowdart dip guaranteed to make hunting small jungle mammals a breeze.

Alive, the frogs, which cost from $35 to $200, require a lot of attention, Bagnall warns. "They can't take extremes in temperature or dryness, and their diet is restricted to very small insects. In fact, you have to raise fruit flies for them." Most of us don't have time to raise fruit flies for our families, let alone for a pet. But, paradoxically, though exotic pet owners are getting busier, they're also getting savvier and more dedicated.

"The whole pet industry has changed," says Bagnall. "Exotic pet owners can't afford to be ignorant, because they're paying more." Making a fatal mistake with a $3,500 miniature ram and ewe is a whole different thing from accidentally offing a twenty-five-cent baby turtle. (I'd like to take this opportunity to make a public confession. I'm sorry, Shelly. I was only seven, and I didn't know that painting your back with nail polish would kill you. Forgive me, too, for digging you up two weeks after the funeral, but I was curious to see if the rumors I'd heard about deterioration were true. You didn't disappoint.) Continues Bagnall, "I can't speak for birds and mammals, but the prices of even standard, bread-and-butter reptiles—boas, garter snakes, pythons—have tripled over the last three years because of government regulation of imports." But the high prices seem to add to the mystique, he says. "Reptiles attract people who want something not everybody has. Also, if you're allergic to fur, they're a nice alternative." (And probably a certain percentage of newly minted MBAs are even now saying to their mentors, "Rep *ties*? I thought you said 'invest in some rep*tiles*!' ") Bagnall adds that poison arrow frogs and Old World chameleons are especially popular now because they've only recently appeared in zoos. "And if a reptile shows up in a movie, its popularity increases tenfold, like the Burmese python in *Raiders of the Lost Ark.*"

Yes, it's a cachet as cachet can world. Perhaps all human progress stems from the tension between two basic drives: to have just what everyone else has and to have what no one has. Covet your neighbor's ass? Get yourself a miniature one and watch him mewl with envy. But be careful: Once an odd animal enters the mainstream, those on the cutting edge of the pet thing have to push for a new personal beast. "Pygmy goats used to be really rare," Fredericka Wagner says with a sigh. "Now everyone has them." (Haven't you noticed that the first question you're asked at the best restaurants these days is "May I check your goat?")

The proliferation of peculiar pets may necessitate a revamping of terminology, says veterinarian Quesenberry. "The term *exotics* is no longer valid. We're talking about animals that haven't historically been domesticated, but they're not wild anymore, either, because they're being bred in captivity and exposed to humans from an early age. Somebody has suggested using the term *special species* for these animals, and reserving the term *exotics* for the zoo stuff."

If you're not ready to pay big bucks for little pigs, you'll be happy to know that the classic exotic pet, the simple yet eloquent sea monkey, retails for just $3.99. Remember sea monkeys? When some of us were kids, during what scientists call the Late Cleaver-Brady Epoch, sea monkeys were advertised in the backs of magazines, usually between the Mark Eden Bust Developer and the Can You Draw This Elf School of Art. A smiling, bikini-clad creature with the head of a monkey and the body of a seahorse promised the requisite hours of fun for kids from eight to—if I recall the stats correctly—eighty. Remember your disappointment when the "monkeys" turned out to be brine shrimp, so infinitesimal that they could only be clearly seen with the enclosed magnifying glass? Remember how not one of them wore a bathing suit? Well, for the same low price, a new generation of kids can learn a powerful lesson about the true nature of existence (sometimes when you expect a bikini-clad aquatic primate, you get a bunch of stupid, skinny-dipping germs). And sea monkeys are easier to house-train than cats.

A Roach Called Asa

□ ■ □

"Sniff this roach!"

Alice Gray cheerfully waves three nauseating inches of creature in my face. I can, of course, refuse, but that seems impolite and unevolved. Gray is, after all, doyenne of cockroaches at the American Museum of Natural History in New York; for over fifty years she's been its chief liaison with the insect world. No doubt there's a good reason for this demonstration.

"Meet *Leucophaea maderae,* a house roach from the Caribbean. See its broad shoulders and tapered hips. It has the prettiest shape of all my roaches." I take a whiff of the lovely and talented *L. maderae.* "And the worst smell! Acrid, isn't it? All insects communicate with chemical messages that say various things: 'Here I am, girls,' or 'Strangers keep out.' "

I'm no entomologist, but today's message seems to be "I stink." Declares Gray merrily, "When I pass a building being torn down, I can tell by the smell which species of cockroach lived there."

Gray's third-floor alcove in a brownstone turret of the museum, beyond the reptile hall and overlooking Central Park, is rich in roaches. On shelves behind her desk are half a dozen ten-gallon glass tanks and five or six two-quart widemouthed jars containing, at the moment, ten species of mostly exotic tropical roaches—a population of perhaps 2,000—plus a few small tanks of tarantulas and a large one holding the chirping crickets they eat. Although she officially retired six years ago from her position as an education specialist, Gray comes in three times a week to tend her collection and discharge her duties as president of the Friends of the Origami Center of America, headquartered in this large tower room. Her colleagues don't seem to mind that the office is bugged. One of them serenely folds a brontosaurus out of purple paper—it's approximately one-zillionth life-size—while I stick my nose into Gray's business.

Gray returns the Madeira roach to the jar it shares with several reeking relatives, then reaches into another full of immature Madagascar hissing roaches. "I can't keep these as wet as I'd like," she says ruefully, grabbing a handful as if they were living licorice. "Hey!" she says to an especially large piece. "You shouldn't be here! This is an American sewer roach, a water bug. He belongs in another jar." She quickly transfers him. "And this fellow belongs over here. This is an ashy gray roach, also called a lobster roach. I have a strange suspicion that if I boiled it, it would turn bright red!"

Gray explains that even though she coats the insides of the jars with Vaseline to keep the roaches in and covers the tops with screening to keep predators like mice out, some jar hopping and outright escapes still take place. "I probably have fifty or sixty loose roaches," she says, "but they don't leave the room." What a relief. "The climate's too harsh outside."

A small, round woman who's been slowed to slightly subsonic speed since heart surgery last January, Gray thinks she's seventy-three or so. "I'm terrible with dates. I can't remember my birthday unless I think of my suitcase—I use it as the lock combination. But I don't recall whether I use the month and day, or the year." She does, however, remember precisely when she came to the museum, fresh out of college.

"It was in October of 1937. I was so pathologically shy that I couldn't have worked in a public capacity; the museum was like a sheltered workshop for me. My parents and I—my father was an engineer who manufactured radios and later microfilm; my mother, a farmer's daughter who kept my father's books—decided when I was in high school on a career in insects; there were jobs for entomologists.

"I've always kept insects as pets, since I was a child of four or five. One of my earliest memories is the smell of potato foliage being chomped by beetles in our yard in Buffalo, New York." Gray's mother had a rule about insect pets that may have turned her daughter into a scientist: "If you can't find out what it eats by dinnertime, she said, you have to let it go. So I found out."

As a teenager, Gray wrote to Frank Lutz, chairman of what the museum then called its insects and spiders department, asking where she should go to college and what she should study in order to work under him. He offered advice and a job. "I think what won him over," Gray says, "was my telling him that I wasn't afraid of hard work and, prophetically, that I was extremely unlikely to get married and leave him."

Lutz suggested she go to Cornell and study not only biology and entomology but also public speaking, scientific illustration, and feature-story writing. "I'm an old-fashioned general naturalist," Gray says. "Today people know more and more about less and less until they know everything about nothing." Later, after she'd been with the museum twenty years, Gray was granted a leave to work on a doctorate in entomology at the University of California at Berkeley. Before she could finish her thesis (on Protura, tiny, transparent insects that live in the soil), Gray was called back to New York to create the insect exhibits for a new Hall of Invertebrates.

But her first task at the museum was dusting specimen cases that hadn't been cleaned in thirty years, by Gray's calculations. She moved on to scientific illustration, then to making three-dimensional beeswax models of insects. "Then I graduated to

PR, answering the public's questions: 'I found something with six legs in the bathtub—what is it?' " (She could often immediately rule out the Andrews Sisters.)

"And about that time, nearly fifty years ago, I made an epic discovery: People weren't watching me to see what I was doing wrong; they were watching me to see what I thought of *them.*" The once-reclusive Gray conquered her shyness with a vengeance; now some of her best friends are mammals.

Gray also began taking a dramatic ensemble of bugs around to grade schools; that's when she started collecting exotic roaches and some other insects. "My aim has been to cure people of their dread of insects. I take four or five extremely large roaches, tarantulas, centipedes, or beetles, big enough to see and slow enough to hold, and let every child smell, touch, or listen to the insect while I tell them about it. In more than forty years, I've had only two failures."

Not counting me. "Want to hold a mature *Blaberus giganteus,* the Madagascar hissing roach?" She offers me a sort of armored stretch limo with legs.

Her joy in the order Blattaria is catching; I really do have a new appreciation of the roach physique, and I'm glad to have smelled one up close and personal. But I decline going mano a mano a mano a mano a mano a mano a mano with it, adding courteously that I'd rather have a root canal without anesthesia. (I've since changed my mind; see "The Novocaine Mutiny," page 90.) Gray chuckles sympathetically.

"People don't like roaches because they smell and scuttle; they're associated with bad housekeeping, and they're thought to spread disease. A pair of misconceptions. Roaches' habits are such that they ought to be prime infectors, but they aren't, and nobody knows why. The American cockroach spends its days in the sewer, then comes up to the kitchen and defecates. If you culture their feces, you can grow microbes that make lab mice sick—but in nature they don't make us sick. Roach allergies are common, however. Their discarded skins grow brittle, crumble, and mingle with house dust.

"I want people to see that roaches are gorgeous! In the tropics—I went to Trinidad two weeks a year for five years with the Junior Entomological Society—roaches think they're butterflies. They fly in the daytime, they visit flowers—they're beautiful!

"I've always been fond of roaches, perhaps because they're the underdogs of the insect world. There are only 3,500 kinds—90 percent tropical—and that's a small number for insects. And their antiquity drew me to them. The most wonderful thing about them is that they've changed so little after 360 million years; fossilized roaches are nearly identical to modern ones. The roach's specialty is lack of specialization. It's so successful because it's not tied to one diet or environment."

Although Gray still occasionally visits local schools and nature centers, she says that these days she spends more time on origami, another longtime passion. "I once picked up an origami book because there was an insect on the cover. Eventually, I made everything in the book." Then one day she was chaperoning a spider appearing on a horror show in the early days of television. She sat in the wings and folded paper, and a stagehand said, "You should have been here last week—we had the origami lady."

"I asked her name, and it was Lillian Oppenheimer, widow of two millionaires and the first origami teacher of note in this country. I called her, and she invited me to see her collection—tens of thousands of pieces. I offered to devise a system of classification, based on the classification of insects. The collection is now here at the museum. Then the editor of the *Origamian,* a newsletter for paper folders, was deported for being a pornographer, and the illustrator went off to be a knife thrower in a circus. I said I'd pinch hit, and I did, for twenty years."

A messenger delivers a pair of small packages, which Gray eagerly opens. "I collect jewelry with an insect motif," she explains, "and I sent some for repairs. I also collect insect fabric, scarves, handkerchieves, and toys—they tell me how the layman feels about insects." She unpacks a heavy silver Navajo bracelet a century old, made in the form of a tiny spider web holding its tenant and a minuscule fly. Gray points out that the fly's wings are on backwards. "Someday," she says, "Betty Faber will get all my insect things, and the roach collection, too."

I meet Faber at a restaurant halfway between the museum and her voice lesson. An entomologist in her late thirties who now teaches biology at a New Jersey prep school, Faber is also a lyric soprano who's sung professionally. For ten years, from 1976 to 1986, she was a research associate at the museum, running the world's largest roach motel.

"I came to the museum because I'd done my Ph.D. work at Rutgers on the feeding behavior of roaches—I'd wanted to conquer my fear of roaches, and there had been an assistantship available to study them. I wanted to meet Alice because I'd heard she had fabulous exotic roaches. All I'd seen at the time were a couple of pest species. She likes to help young people, and we both liked cockroaches; we spent many happy hours discussing entomological matters. One of the best things she did for me was suggest me as her replacement when she couldn't lead Trinidad trips anymore. I'm lucky to have such a marvelous friend."

Faber soon found her niche at the museum. "There was a 40-by-60-foot greenhouse on the roof of the building that was infested with American cockroaches; they hadn't sprayed it because fish were kept there. I was invited to study those roaches."

No one had studied roaches on the loose before, so Faber didn't know what to expect. "First I labeled all the roaches with medical adhesive tape and wrote their names on it: Asa, Bea, Lee, T.J. But there were so many roaches I quickly ran out of names with three or four letters. Numbers turned out to be the best way to do it. In the winter there might be as few as ten or twelve labeled adults and a lot of young roaches, but in the summer there might be two or three hundred adults."

Before Faber's research, entomologists thought that roaches were active both in the evening and in the early morning. She discovered, however, that although roaches are active in the early part of the evening, they start disappearing at around one or two in the morning.

In the lab, Faber says, cockroaches are kept in a contraption that resembles a Ferris wheel. Activity in the wheel clicks on an electric switch that lets the investigator know when the joint is jumping without watching twenty-four hours a day. "I think that in the lab, where lights are set on a timer to duplicate the length of a natural day, roaches seem to be active in the evening *and* early morning because they get scared when the lights go on, and since they can't hide, they just keep running."

Males tend to stay out later at night than females, Faber found. "People always read a lot into that," she says. By four in the morning, even the guys are gone. "So if you get up at night to get a Coke, you stand a better chance of stepping on a cockroach early in the evening than late in the evening, for whatever that's worth."

Making the world safe for late-night snacks isn't what motivates Faber. "There's nothing like being the first person to know something. It's exhilarating. I love science fiction, but when I'm doing my roach work, I don't feel the need to read it; then my life becomes its own science fiction. I can't wait for the next chapter to unfold."

But three years ago, shortly before Faber had her son, Josh, the museum's animal behavior department was eliminated and the rickety rooftop roach motel condemned. When her work is published and Josh is a little older, Faber says, she'll concentrate on building up Gray's exotic roach collection for further behavioral studies and elementary school road-show-and-tell. "I've got some at home, and I can get some in Trinidad. And there are various people around the city with colonies who I can call on at any time." I can think of eight million or so who'd be happy to have Faber drop by to borrow a cup of cockroaches.

The message she and Gray want to spread, Faber says, is that learning to appreciate any living thing is enhancing. "Even roaches, which people consider loathsome but which are beautifully designed, streamlined, interesting creatures.

"It's not necessary to become a nature freak to understand that we're part of a bigger whole that has nothing to do with buildings or Bloomingdale's. Once you make that shift in point of view, maybe it resonates throughout your life.

"Although I've seen people make some profound shifts in that way and be the same old horrible person in every other way."

For the past six years, Faber's been bringing insects to inner-city schools under the auspices of the New York Academy of Sciences. "It's amazing to watch kids open up to science when they realize that these cockroaches that they live with are actually considered worthy of study. They volunteer a lot of information; they discover they have empirical data that's useful and they start to get confidence. Roaches won't cure the world's ills, but I think they can be a very positive force."

Of Human Bandage

□ ■ □

Now when I die
Now don't think I'm a nut
Don't want no fancy funeral
Just one like old King Tut
—Steve Martin

The Summum Corporation of Salt Lake City, the world's only commercial mummification company, is encountering stiff competition. Although it's had tremendous success with animals and body parts, the company has yet to lay its wrap on an entire human being. Medical schools get first crack at fresh bodies. So Summum is looking for a volunteer.

According to the company's founder and president, Corky Ra, 109 people have signed up to be dipped in secret sauce, covered with plastic, wound with linen, encased in resin, hermetically sealed for eternity in a mummiform of bronze, stainless steel or fiberglass, and displayed in the venue of their choice. But Summum's clientele has failed to fulfill a key requirement for mummification: none has met his Maker, or even exchanged postcards. So far, the company's only demo-mummies—on display in gold-leaf mummiforms at its pyramid-shaped headquarters—are Butch and Oscar, Ra's former Doberman and cat (who were not coerced into the job, you'll be pleased to know, but shuffled off this mortal coil due to old age and feline leukemia, respectively).

"We advertised in some California papers, offering to do it for nothing," says John Chew, head of the funeral sciences department at the College of Boca Raton in Florida and one of two people who perform the actual mummification for Summum. "We're also trying to work through medical examiners and anatomical boards, the people who supply cadavers to hospitals, to find an indigent, someone with no known relations, or relations who say, 'Yes, I would like that.' "

They'd be getting a good deal; mummification can easily cost $35,000 and up. And the folks at Summum rave about their product; esprit de corpse is high. "It's phenomenal. In all the tests we've done, nothing has decomposed," says Chew's fellow mummymaker, Ron Temu. "We checked out Butch after a couple of years and he's absolutely perfect. You can still move the skin on his back, and his eyes look real good."

Ra says he first perceived the need for a commercial mummification program in the mid-seventies. "The funeral industry was allowing people to believe that when they're embalmed, the body is preserved. But that isn't true. You spend lots of money for a casket, and then you decompose."

That really fried Ra's hide. He began studying body preservation techniques of the ancient Egyptians, whose afterlifestyle depended on an undecayed body in which the soul could dwell. "I saw that it was scientifically possible to recreate and improve on those methods," says Ra (formerly Nowell; he legally renamed himself after the ancient Egyptian god of cheerleading nine years ago because he thought it would be good for business). Ra met Chew, a self-described "Egyptian bug" at a funeral directors' meeting and discovered they shared a fascination with certain ancient funerary customs. (Just how does that come up in conversation?)

"Since 1979," Ra says, "we've done exhaustive laboratory and field tests, experimenting with how different types of salts react to cellular structures—we started working with chicken eggs, because they're single cells—and arrived at our special formula. We have patents on the actual treating of the body and the containers of bronze or stainless steel, called mummiforms. And the word *mummification* is our registered service mark."

Tut-tut, you say; impossible. Someone must have beaten him to the registration desk by 4,500 years or so. But the word—from the Arabic *mumiya,* meaning bitumen, used in the embalming process—is apparently his, pharaoh and square. "We're registered in all fifty states," says Ra, who studied business at the University of Utah but has since become a licensed funeral director. He also runs half of Utah's wineries (the state has two) and teaches aerobics on the side.

Mummies-to-be may choose from three standard mummiform styles—Egyptian (with or without personalized hieroglyphics), art deco, or Renaissance—featuring, if they like, a mask cast from their living face. Or they can get a custom job, carved with the design or outfit of their choice. Ra has taken orders for military uniforms (he keeps track of any new decorations earned by the predeceased), three-piece business suits, and a gown with cape. One customer has asked that inside the mummiform he be wrapped in silk embroidered with verses from the *Tibetan Book of the Dead.* Another, an avid do-it-yourselfer, plans to travel to the great beyond accompanied by his favorite wrench.

Temu has his mummiform all planned: "It's going to be very plain, very smooth, very modern looking; almost a capsule, but with a slightly defined body and my face—sort of how an Oscar looks." (That's Oscar the Academy Award statuette, not Oscar, Ra's cat. Maybe you'd prefer to be a Mummy who's a Grammy. For best wrap music, I guess.)

You're probably wondering how Summum's mummies will stack up against the Egyptian originals, some of which have had a shelf life of several millennia. "The Egyptians dehydrated and thus preserved the body with natron, a naturally occurring mineral salt," says Chew, "but we use a wet method. If we dehydrated the body, it wouldn't be viewable at a traditional funeral. The body will be embalmed so there can be a funeral service, then shipped to us in Boca Raton, where we'll place it in a stainless steel vat of fluid that's a combination of salts, oils, alcohol, and other chemicals, as well as natural substances similar to the aloe plant, that inactivate the tissues. The body breaks down through two processes—its own chemistry, powered by oxygen, and bacteria. By saturating and inactivating every cell, driving out the oxygen, and replacing it with our formula, we've eliminated both processes."

The Egyptian method took about seventy days, according to the accounts of the Greek historian Herodotus, and came in three price ranges: deluxe, regular, and no-frills. In a top-of-the-line mummification, the heart and kidney stayed put and the brain was removed through the nostrils—to this day a great party stunt, second only to doffing your vest without taking off your jacket. Other organs were removed through an abdominal incision, dried, wrapped in linen, and placed in special jars, each sacred to one of the god Horus's sons: Imseti, in charge of the liver, Duamutef (stomach), Qebehsenuef (intestines), and Hapy (lungs). Why no jars honored sons Snezy (nose hairs) and Dopy (little flecks of saliva in the corner of the mouth) remains a mystery. The body was then filled with resin or sawdust, dried in natron, anointed with oil and "every kind of spicery except frankincense," according to Herodotus, and wrapped in strips of linen. The historian added that the middle-range mummy was injected with cedar oil instead of having its innards removed, and the economy job was simply cured, with no courtesy wrap.

The Summum method preserves the internal organs in situ, except for the brain, which is removed through an opening in the skull, embalmed, then replaced. After the vatting, which can take up to thirty days, depending on the size of the body, the mummy-in-progress has a rubbery consistency but gets firmer as it dries, Chew says. He and Temu smear it with a special lotion that leaves the skin soft and supple, then cover it before it's completely dry with a patented polyurethane seal something like a large surgical glove. Then, looking, one would imagine, like a mammoth leftover, the body is wrapped in linen, cocooned in a resin cast, and placed inside a mummiform. "We evacuate all the air," adds Temu, "replace it with inert gas, and weld the mummiform shut with another patented process. As long as no one opens it, nothing can corrupt the body."

Temu would like to emphasize that mummification is a one-way ticket. "We can preserve the body's genetic message so people may someday have themselves cloned." Scientists at the University of Uppsala, Sweden, have reproduced segments of DNA taken from a 2,400-year-old mummy. "But we have to make it clear that this isn't like cryogenics; they can't be reanimated. This is simply an option for people who don't want to allow their bodies to fall into corruption after death."

Personally, I want my body to fall into corruption *before* death, while I can still enjoy it. But Janet Greco, a thirty-seven-year-old Salt Lake City nurse who, along with her husband Al, signed up at Summum four years ago, agrees with Temu. "I'm not an egotistical person, but I have a respect for my body," says Greco. "I work out three or four times a week and eat healthy foods. It doesn't make sense to spend all this time taking care of my body and then to discard it." A friend who knew how she felt told her about Summum. "We talked to Mr. Ra, then it took about a year to make a decision." (Al subsequently took a job at Ra's winery.)

Like all of Summum's customers, the Grecos each bought an insurance policy that will cover mummification. Don't bother trying to track down the bandage division of Mutual of Heliopolis. "Different people use different companies," Ra says. "You get what's called a universal life policy. If you're in your twenties, you can make a one-time payment of perhaps a couple of thousand dollars. If you're older, it's more expensive. You have to plan ahead. If you needed it tomorrow and you were going to pay cash, the actual mummification process would cost $7,700 and mummiforms start at $26,000." (In ancient Egypt, according to the Greek Diodorus, first roll-on historian, the deluxe wrap cost one silver talent, the standard twenty minae, and the economy $19.95 plus postage and handling.)

Greco doubled the life insurance policy offered by her employer. "I pay a few dollars extra, but it's worth the peace of mind that comes with knowing that everything's planned and I'm ready to go.

"My mummiform will be very simple, with a life mask. I may take a few little things with sentimental value—some letters from my husband and family, my wedding band, and a necklace. I'm not sure where I'll be. Mr. Ra is thinking of opening a mausoleum in a mountain range south of Salt Lake City and I think I'd like something like that," she says cryptically. One of Summum's brochures (which range from sober discussions aimed at the funeral industry to some fairly zany stuff about Atlantis and "cosmic makeovers") discusses Ra's proposed condos made of stone-a. It sensibly points out that "earthquakes can level and destroy all normal mausoleums. . . . Vandals may pillage. A riot or revolt, caused by political agitators may lead to burning of cities." Summum will offer tamper-proof sepulchers carved into a granite mountainside.

The company gets several inquiries a day from around the world, Ra says; so far, he's heard from a highly placed official in the principality of Andorra, an Indian government minister, several South American heads of state, the funeral division of the U.S. Army, and, of course, several film and music celebrities. Ra is so convinced mummification will become the way to go that he plans to allow funeral homes whose personnel attend a special training program to purchase mummification franchises. Like fast-food places, they'll have to buy their equipment—soaking vat, chemicals, wrappings, resins—from the parent company.

Maybe soon a chain of golden sarcophagi—"over three billion swathed"—will dot the nation. Maybe someone will decide to kill (and preserve for eternity) two birds with one stone, opening establishments that offer both physical and spiritual refreshment, called, perhaps, Coke 'n' Croak. Maybe you'll volunteer to be Ra's first subject. Or maybe you'll just say yo' mummy.

Lookin' for Science in All the Wrong Places

□ ■ □

Anthropologist James Schaefer has spent more than a decade studying elbow-bending rituals of the Western Honkytonkin, a fierce people recognizable by their characteristic red necks, white socks, and Blue Ribbon beer. He's discovered that the fancifully titled hymns accompanying their worship of spirits—incantations like "She's Acting Single, I'm Drinking Doubles" and "I'm the Only Hell Mama Ever Raised"—have dangerous powers. To hear him talk, you'd think jukeboxes should come with warnings from the Surgeon General.

Schaefer, now director of the Office of Alcohol and Other Drug Abuse Prevention at the University of Minnesota, conducted two long-term studies of the effects of barroom environment—music, lighting, decor—on the drinking behavior of some 5,500 country-and-western bar patrons in Montana and Minnesota. He found, among other things, that country music is more likely to lead to alcohol abuse than are pop or rock. And the slower the tune, Schaefer says, the faster the drinking.

That wasn't his hypothesis when he first began observing the patrons of honky-tonks and juke joints in Missoula. "I'd expected the opposite—that when the beat started cranking, they'd get to drinking." Instead, he was surprised to learn there was a direct blues-to-booze correlation. The songs that most often plowed 'em under were Kenny Rogers's "Lucille" and Crystal Gayle's "Don't It Make My Brown Eyes Blue." Other singers likely to up the tears-to-beers ratio were Willie "Blue Eyes Cryin' in the Rain" Nelson, Hank "So Lonesome I Could Cry" Williams, Waylon "You Can Have Her" Jennings, and Merle "Misery and Gin" Haggard.

The idea of working in close collaboration with Dr. Haggard and Professor Jack Daniels came to Schaefer one night at the Trail's End Bar. "I'd just started teaching at the University of Montana where I'd been an undergraduate. I'd finished my doctoral work, a cross-cultural survey of drinking patterns worldwide, and I had a grant to study alcohol metabolism among Native Americans.

"I decided to revisit one of my favorite bars from the old days. That night everything was really moving, until the jukebox suddenly stopped. The place died. Sammy Thompson, the bartender, passed the bean game around—whoever draws the red bean out of a jar has to pay for the jukebox. The loser got out his quarters and the crowd started shouting out favorites—G12! D6! The soundtrack had stopped and people wanted the mood back. Looking at my old haunt with an anthropologist's eyes, I saw that the energy and interactions in the bar were conditioned by the music." A better name for jukebox, he says, would be "mood-selection device."

Schaefer decided right then and there to study formally the effect of music on drinking in country-and-western bars. "A graduate student, Paul Bach, came up with a very elegant experiment. We took a tape recorder and 120 minutes of tape and sat down at a table in front of the jukebox; our field of observation included about 30 percent of the bar. Every time someone sipped a drink, we tapped the table. We did this in several bars, with the music in the background. Later, we ran a correlation of sips per beats of music. Eighty-five beats a minute and above led to the least drinking. A tempo of 60 beats a minute led to excessive drinking." Heartaches by the number, troubles by the threescore.

Other social scientists have speculated that the reason downbeat and down-home combine to boost boozing has something to do with slow music's effect on heart rate or the autonomic nervous system. But Schaefer thinks content is a co-conspirator. "We looked at the lyrics of songs that stimulated drinking, and it turned out they were the tearjerk lonesome blues, songs about the sad and often abusive aspects of life among certain working-class people. Hard drinkers were more likely to choose slower-paced, wailing, self-pitying music."

Schaefer thinks that drinkers in bars playing rock, on the other hand, are chiefly affected by rhythm, not lyrics. "Rock is more sterile and repetitive," he says. "Country has a story line that hooks into a core value system. The lyrics help people rethink their lives." (Consider, for example, David Frizzell's "I'm Gonna Get a Wino to Decorate Our Home," or Loretta Lynn's "Your Squaw's On the Warpath": "That firewater you've been drinkin'/Makes you feel bigger, but, chief, you're shrinkin'. . . .") Schaefer considers rock themes more callow and shallow. (Just compare the briefly popular rock anthem "Boom, Boom, Let's Go Back to My Room," by Paul Lekakis, with the tender country ballad "Get Your Biscuits in the Oven and Your Buns into Bed," by Kinky Friedman.)

Grand Old Soap Opry lyrics and lugubrious tempos aren't the only factors that make honky-tonk habitués nip at what singer George Strait calls "that 80-proof bottle of tearstopper." It don't mean a thang if it ain't got that twang. Fans get drunk on whine: Country singers wail from the heart and through the nose. Schaefer cites the work of pioneering ethnomusicologist Alan Lomax, who determined that nasal music is found in societies with a great deal of tension. Tribal warfare, for example, is correlated with nasality. (How a few airlifts of Dristan

Nasal Spray might have changed world history!) "It's a leap from Lomax's cross-cultural data to nasal country songs, but it's interesting to me that there's tension and anxiety being explored in the songs of some of the best twangy country singers like George Jones, John Conlee, and Merle."

Schaefer has identified a number of variables that influence the amount of drinking in a bar. A strict, posted dress code and an even man-woman ratio contribute to moderation. Heavy tippling goes with dim lights, cheap drinks, and a small dance floor. A live band makes the liquor flow more swiftly than a jukebox does. Even decorative art makes a difference—landscapes tend to slow drinking rates, while repeated patterns, portraits, and especially action photography are escalators. Lord only knows the dire effects of those revolving lamps printed to resemble real-life shimmering falls in the land of sky-blue waters.

Yeah, it sort of looks as if Schaefer dreamed up a research project that let him hang out in bars all night listening to "She Got the Gold Mine, I Got the Shaft" while donating his liver to science. Let it be known that he hoisted only the minimum number of brews necessary to keep from blowing his cover. And it wasn't all skittles and beer; the assignment was more dangerous than it sounds. We now know that songs about losin', hurtin', cheatin', workin', and dyin' lead to drinkin'. And drinkin' leads to fightin'. (Not to mention tryin' to keep from trippin' over all those dropped *g*'s litterin' the floor.) If patrons had caught Schaefer and Bach studyin' them, they would have been sorely displeased. "There's a lot of pressure to conform at a country-and-western bar," Schaefer says. "You've got to wear the right kind of belt buckle and tip your hat at the right rowdy angle." Country music may be unsafe at any speed. "When a fast song came on, there was a hoot and a holler and the place erupted into what was almost a drunken brawl—lots of backslapping and hat waving. If you didn't join in, you were a goner. There were times I had to participate fully in the complete episode." For you nonanthropologists, that means Dr. Schaefer drank and danced.

Maybe you're thinking that someone should write a song about Schaefer's findings. Well, someone did. After hearing Schaefer lecture, playwright Charles Rothaus urged him to contribute a song to the musical Rothaus was writing about alcoholism. Schaefer's tune, "Slow Country Music," became part of *Easy Does the Stars,* which had a respectable 1983 run in Austin and Atlanta. "Joe, don't play that slow country music/I drink more and think sore and sing right along/I need a dose of dancing close, but the girls are all gone/Spare me the pain of a slow sad song." I think it's damn good, and if you don't, we can just step outside. I'd like to see Willie Nelson publish a paper like Schaefer's "Alcohol Metabolism and Sensitivity Reactions Among the Reddis of South India"!

Schaefer's song got me to thinking. Excuse me, thinkin'. Perhaps more anthropologists should follow his lead, compressing the essence of their often dryly reported work into musical abstracts in the C&W mold. Anthropology is, after all, the study of mankind, and what is country music but the cry of the common man and woman? Wouldn't you be more likely to read this article from the *Journal of the American Anthropological Society*—"The Effects of Geochemical Factors on Prevalence of Dental Diseases for Prehistoric Inhabitants of the State of Missouri"—if it were called "Holey Molars and Holy Rollers Have Run in Our Family for Years, Dear, Thanks to a Millennium of Selenium"? Margaret Mead's *Coming of Age in Samoa* is actually rather juicy, but wouldn't it be even more fun to *sing* her landmark study of free and easy adolescent sexuality among early-twentieth-century islanders? "You said you agreed/With old Margaret Mead/That sexual expression /Was better than repression/You swore to meet me/Tonight 'neath the palms,/Unwrap my lavalava, lover/Hold me in your arms/But now your new taboo/Is causin' me taboo-hoo/Since you kicked my butt/Right out of the hut."

In fact, maybe all scientific disciplines should go country. Consider a musical exploring the possibility that researcher Rosalind Franklin unlocked the secret spiral of DNA before Francis Crick and James Watson, but was cheated out of credit for her discovery. A somebody-done-somebody-wrong song could spell out just how Watson might have screwed her: "Let me into the lab tonight Rosie/In fact, let me into your genes /Your X won't know Y though he's nosy/And you'll know what happiness means/Promise you'll meet me at seven/We'll climb the double helix to heaven/Then I'll pilfer your files while you gaze in my eyes/Lord willin' and Crick don't rise."

How about a double album called "Principia! Juice Newton Sings Isaac Newton"? And what about the humanities? Wouldn't you pat your feet to a critical study of *Death in Venice* called "Stand By Your Mann?" I myself have composed a piece on interfaith/intercultural marriages called "Shikse from Dixie."

These days Schaefer's moving on to bigger and wetter things. "The Montana work was good social science but not good quantitative research. When I arrived at the University of Minnesota, my colleague Richard Sykes and I put together a scientifically objective study of sixty-five bars, three of them country and western. We trained our observers to look at the drinking practices and self-regulatory patterns of naturally occurring social groups." Two striking findings have emerged from the ongoing study, funded by the National Institute of Alcohol and Alcohol Abuse.

"First, the one big predictor of alcohol abuse was time. The shorter the stay in the bar, the faster the rate of consumption and the greater the risk of abuse. Singles—54 percent of the people observed—stayed the shortest time and drank the most, an average of four drinks in fifty-six minutes. We consider alcohol abuse to be drinking at a rate of three or more drinks an hour. At a certain point, the rate per hour of people in foursomes through ninesomes slowed down. The group influence appears to be a powerful microculture."

In the next phase of the study, Schaefer and colleagues will try to determine how groups make decisions about drinking. They'll recruit 200 groups of drinkers who'll be wired with voice-activated microcassettes; later, computers will sample twenty-minute segments of conversation. Trained observers will also be watching. "We want to know what happens when the glasses are empty. Who decides to reorder and why? Do they reorder when the waitress is near, or do they insist on her coming over? If someone says, 'I'm gonna dance with that good lookin' lady over there,' do they still include him in reordering? At closing time, when the bartender shouts, 'You don't have to go home but you can't stay here,' do people have a last round, and do they urge the driver to have a cup of coffee?"

Schaefer knows his work sounds kind of goofy. But identifying environmental variables that trigger alcohol abuse—from the face on the barroom wall to "Thank God and Greyhound (She's Gone)"—means being able to manipulate those variables and thus cut down drinking. "We need to educate patrons and bar owners about what constitutes a risk so people can drink smarter," Schaefer says. "Looking at drinking patterns can help us plan an alcohol diet, like a food diet, and to create concrete measures of accountability in bars." His work has already led to the creation of prevention programs involving bar staff. "But drinking is still killing a lot of people. And last year alcohol abuse cost this country $120 billion, 75 percent from lost productivity due to absence and impairment."

See? Schaefer's no candidate for a Golden Fleece Award. He's doing important work. Someone should endow a chair, or maybe a stool, in his honor. I think he deserves a floral tribute—maybe four roses? Jim Schaefer, this bud's for you.

I'm Okay, You're History

□ ■ □

Among the tribal communities of Kohistan, a region of Pakistan near the Afghan border, if a man flashes light from a mirror at another man's wife, he is dealt instant blood revenge, or *mar dushmani.* Even if it was a mistake—"no reflection on your wife, heh-heh"—he's dead meat.

I'm sorry I didn't learn about *mar dushmani* until after my recent graduation from Don't Get Mad . . . Get Even, a crash course in vengeance offered by the Discovery Center, a New York City alternative adult education program. I'm sure my classmates would have taken detailed notes.

I went mostly to see who'd be enticed by a catalogue description that read, "Has someone done you wrong? . . . Are you sick and tired of being pushed around? Vengeance is quite fulfilling, and very healthy, too." My twenty-seven classmates seemed normal enough; none looked like the type anxious to impress Jodie Foster. The investment banker sitting next to me explained that he'd signed up because he'd been shafted one time too many. "People who give service just don't care anymore. They have no pride or humanity. It's infuriating. You can't kill everyone," he said philosophically, "so I thought I'd get some advice on how to handle them."

Our instructor, Alan Abel, a Peter Ustinov look-alike in his mid-fifties, announced that he's an amateur avenger, a marketing consultant, and the screenwriter of the 1974 movie *Is There*

Sex After Death? He ought to know. A rather successful hoaxer, he faked a disappearance in 1980 and had the rare pleasure of reading his own obituary in the *New York Times.*

Abel asked us to write down stuff we were sore about. I mentioned a new neighbor who plays his stereo so loud that an irate couple in Beijing bangs on their ceiling. From what I could see, bosses, landlords, and exes figured heavily in other people's grievances.

But the lists were for later. The first half of the class dealt more with the preemptive than the retaliatory. (And that's perfectly fine. I'd never presume to criticize an expert in revenge. Abel is a very nice guy. And, in case they're reading this, I'd like to mention that the Corleone boys are also very nice guys.) Abel offered tips on how to get three seats to yourself on an airplane—dangle a piece of string from your mouth; it scares people—and how to get a table at a crowded restaurant by pretending to be a doctor. A lot of his tips require props, like string and a stethoscope.

After the break we got into hard-core tit for tat. Abel told us how he got back at the Helmsley Palace Hotel in New York for barring him from the men's room because he wasn't a registered guest. (He rented a portable outhouse and planted it on the sidewalk in front of the hotel with a sign that said For Unregistered Guests. The hotel management threatened to call the johndarmes but backed down. And Abel recently got the last laugh when queen and felon Leona Helmsley was herself dethroned.) He also recounted what happened when a famous talk-show host refused to pay some money she owed him for a guest appearance. (He got friends from around the country to mail her small amounts of cash accompanied by a sympathetic note saying they'd heard Abel talking about her on the radio and were sorry she couldn't pay her bills.)

One man in the class wanted to know how to get even with a woman who had jilted him. Abel suggested checking into a hospital for tests and having friends make her think he was at death's door so she'd feel guilty. (Which is fine for folks whose health insurance policies include unrequited love coverage, but pricey for the rest of us.) We never did get to the stereo issue, although after class a college student gave me the name of a book about making bombs at home, and someone else suggested hiring a mariachi band to play outside my neighbor's window.

After this introduction to the practical aspects of revenge, I wanted to know more. Sure, I was familiar with its popular manifestations (from short-sheeting to cement overshoes), with conflicting biblical injunctions (so, which works—the Old Testament's eye for an eye or the New Testament's turn the other cheek?) and with the classic revenges, like Medea's and Montezuma's. And I knew that living well was the best. But what did *Science* have to say?

Science was away on a ski trip, so I called a fellow who studies organized vengeance for a living. "Our society doesn't have institutionalized vengeance," explained R. Lincoln Keiser, a Wesleyan University anthropologist. "But it's common in stateless societies, those that anthropologists call acephalous—literally, 'headless.' There, social control is often maintained by the principle of self-help: It's up to the individual to enforce the rules. And one moral obligation is retaliation for the death of a kinsman or for other insults to honor." That's true today, Keiser says, of the Kohistani tribes, among whom he lived, and in remote parts of Greece and Albania.

The classic work on organized revenge (besides, of course, the *Congressional Record*) was done by anthropologist E. E. Evans-Pritchard in the 1930s. Studying the chiefless society of the Nuer people of the Sudan, he found that the threat of aggression from a wronged individual and his kinsmen maintained order. "From Evans-Pritchard's work you can develop the theory that, in a sense, the whole world can be considered a stateless society," Keiser says. "The balance of power is maintained by the threat of revenge. And in order for that threat to be real, there must be periodic outbreaks of violence." On a small scale, he says, this theory explains blood feuds; on a larger scale, the wars in Vietnam and Afghanistan. "According to this view, until we change the world order, we're doomed to these periodic outbreaks of violence."

Vengeance is considered a dandy evolutionary strategy by sociobiologists, those who believe that human behavior is chiefly motivated by the drive to perpetuate the species; or, as they like to put it, positively whooping with merriment, "A human being is DNA's way of making more DNA." Anthropologist Napoleon Chagnon, who has been studying the rather curmudgeonly Yanomamo of Venezuela since the 1960s, noted that men quick to take vengeance on their enemies collected the most wives and thus had the most children, earning them the title Most Genetically Successful, if not Mr. Congeniality.

Among Kohistani tribes, the willingness to mete out blood vengeance is part of what makes you an upstanding citizen and jolly good fellow. Keiser once asked a Kohistani friend to arrange a village tour for some visiting American scholars. Later, Keiser casually mentioned to his friend that one of the villagers had come on a little strong to a woman in the group. "Tell me who it is," the man said matter-of-factly, "and I'll go kill him."

Our culture's take on the matter is, obviously, different. Americans certainly have vengeful feelings, Keiser says, but we don't usually act on them because to do so is socially unacceptable.

"Americans go in for third-party revenge," says a friend of mine who considers himself a lay expert on what he calls the modern revenge movement: the fantasy of shattering society's taboo against big-time retribution. A mild-mannered family man and respected professional whom I'll call Conan the Librarian, he's seen and cherished every movie ever made starring Charles Bronson, Clint Eastwood, or Arnold Schwarzenegger: *Death Wish, The Terminator, The Defenestrator, The Osterizer.*

"Look at the tradition of calling on God to destroy our enemies, or at the criminal justice system," Conan says. "I happen to think that today the most American brand of revenge is the nuisance lawsuit—people don't mind the pain and expense of litigation, just so it screws the other guy for a long time. But for a good discussion of the American attitude," Conan continued, "you should look at a book called *Wild Justice: The Evolution of Revenge,* by Susan Jacoby. And rent *Revenge of the Nerds.* It's very satisfying."

Jacoby thinks that revenge shouldn't be treated as a "sick vestige of barbarism" but as an acceptable human emotion too powerful to deny. "Dismissing the legitimate aspects of the human need for retribution," she writes, "only makes us more vulnerable to the illegitimate, murderous, wild impulses that always lie beneath the surface of civilization. . . ." When the social institutions that are supposed to take over the job of vengeance from individuals fail, the result is a rise in the personal vendetta.

Psychiatrist Willard Gaylin, president of the Hastings Center, the bioethics think-and-feel-tank, agrees with Jacoby. "As long as the state is doing the righteous thing, vengeance is controlled. But when people think that goodness doesn't pay or badness isn't punished, it can lead to vigilantism."

I'd discovered what Science says about revenge. What does it *do* about it? "Scientists are generally humble about seeking revenge," observes George Tremberger, director of the History of Physics Laboratory at Columbia University. Though he can think of many bitter rivalries, Tremberger says, the only revenge he can recall is pretty mild: Humphry Davy, inventor of the miner's safety lamp and a prominent member of the Royal Society in the 1820s, believed—wrongly—that Michael Faraday had stolen the theory of electromagnetic rotation from a friend of his, William Wollaston (who probably kept his theory under the mat on the front porch, the first place any rival would look). To right the imagined wrong, and perhaps to assuage a touch of envy, Davy, a regular Attila the Hun, tried to block Faraday's election to the Royal Society.

Michael Ruse, professor of history and philosophy at the University of Guelph, Ontario, doesn't think that vengeance among scientists is always so genteel, although it rarely gets more vehement than a public riposte. Consider the bitter fight between nineteenth-century comparative anatomists T. H. Huxley and Richard Owen—ostensibly a fight about evolution, but actually a personality clash. Both originally thought the theory of evolution was a great idea, but after Darwin made Huxley his protégé and mouthpiece, Owen was so miffed that he took an opposing position. "He was torn," Ruse says, "between saying 'this is bull' and 'I thought of it first.' " To get back at Huxley, Owen sicced on him the popular and grandiloquent Bishop Samuel Wilberforce of Oxford during the famous open debate on evolution at the 1860 meeting of the British Association for the Advancement of Science. Wilberforce brought down the house by asking Huxley whether he was descended from a monkey on his grandmother's or his grandfather's side, and Huxley, himself a righteous razzer, laid them in the aisles by retorting that he'd rather be descended from monkeys than from a man "who—not content with an equivocal success in his own sphere of activity, plunges into scientific questions with which he has no real acquaintance, only to obscure them by an aimless rhetoric, and distract the attention of his hearers from the real point at issue by eloquent digressions and skilled appeals to religious prejudice." (I've got to remember that one the next time a bishop busts my chops.)

A more recent rivalry worth noting, Ruse says, is that between sociobiologist E. O. Wilson and population geneticist

Richard Lewontin, both of Harvard. Lewontin turned against Captain E.O. and pseudonymously attacked his colleague in the British journal *Nature.* "I think," says Ruse, "the attack was Lewontin's way of striking back for what he perceived as accumulated political and ideological wrongs; it was two alpha males struggling to be number one."

Among science students, Caltech may be the capital of retaliation. A particularly delicious incident in the early 1970s involved a math professor who annoyed students by his mechanical, predictable approach to teaching—his lecture notes were straight from his book. One student got hold of a device that changed the normal frequency in an electric outlet to any desired value. He plugged the classroom clock into it and, over several weeks, upped the clock's speed, first by 10 percent, then 12.5 percent, then 15 percent. Each day, the frazzled professor raced through the tired material more and more swiftly, until finally he was reduced to fast-forward gibberish.

Now I think I know what to do about the new neighbor with the stereo. Last night he played six hours of Motley Crüe (or some other group that sounded as if it had an umlaut). If he does it again, I'm seriously thinking of blocking his induction into the Royal Society.

Scents and Sensibility

□ ■ □

Dr. Susan Schiffman, a Duke University psychologist, studies the effects of fragrance on the emotions. She's been hired to come up with an aroma that could be sprayed in New York City's subways to reduce commuter aggression and increase friendliness. Because of a confidentiality agreement, Schiffman can't reveal the identity of her client or the fragrance she thinks will soothe the savage straphanger. I'd say her best bet is chloroform.

"The idea," Schiffman says, "is to use the fact that smell is the most evocative of the senses, because it's so intricately connected to the brain's limbic system, the area that controls emotion."

Although it may be a while before the designations "local" and "express" are replaced by "scented" and "unscented," other projects involving odor engineering—using the link between odors and emotions to predict and manipulate human behavior—are well under way.

"We have already developed a system to control the environment by fragrance," says Junichi Yagi, vice president of S. Technology Center-America, a subsidiary of Shimizu, Japan's largest architectural, engineering, and construction firm. According to Yagi, Shimizu's Aromatherapeutic Environmental Fragrancing System, which delivers behavior-altering scents through air-conditioning ducts, has been found to calm restless nursing home patients and to enhance efficiency and lower stress among factory and office workers.

Experiments in Japan with thirteen keypunch operators, monitored eight hours a day for thirty days, showed that the average number of errors per hour dropped by 21 percent when office air was scented with lavender (it reduces stress) and by 33 percent when laced with jasmine (it induces relaxation); a stimulating lemon scent reduced errors by 54 percent. "The keypunchers enjoyed the fragrance," Yagi says. "Even when the scent was below conscious levels, they reported feeling better than they did without it. And we've learned that you don't have to deliver fragrance all the time." Precise fragrance levels and the optimum delivery schedule are trade secrets, Yagi says.

Shimizu researchers also find that orange, peppermint, eucalyptus, chamomile, and Japanese cypress are soothing, while scarlet sage and rosemary are stimulating. (I would add Play-Doh and Rockin' Roger's Bar-B-Q Sauce to the stimulating list.) The company has received inquiries from American banks interested in energizing their employees; some hotels are considering using fragrance as a sort of olfactory Muzak.

The concept raises many questions. Could, for example, kids taking the SATs or Olympic athletes be disqualified for illicit citrus use? What about the opportunities for sabotage? Industrial spying may be replaced by stench warfare: A sneaky competitor could spike a company's system with enough chamomile to send the whole staff reeling into the arms of Morpheus. And suppose greedy heirs replace granny's soporific jasmine with an overdose of rosemary, causing her to boogie to death? Still, the more I think about it, the more I'm convinced that Lemon Pledge has saved me from some potentially fatal furniture-polishing errors.

Schiffman has experimented with using puffs of aroma to keep assembly-line workers, pilots, and long-distance truckers alert. An odor is invigorating, she says, if it stimulates the trigeminal nerve, one of two nerves in the nose that receive signals from smells. "The olfactory nerve is the one that allows you to tell the difference between oranges and roses. The trigeminal nerve detects irritations, like smelling salts, or temperature, like the cooling effect of menthol. There's a tendency for aromas with a low trigeminal component to calm people and odors with a high trigeminal component to serve as pick-me-ups." That's because trigeminal stimulants, Schiffman has shown, increase blood levels of adrenaline.

Some odors seem to have inherent properties that relax or invigorate, but others affect behavior because of earlier experiences or sensations we might associate with them. Schiffman often deliberately creates such associations to change behavior. "I run the weight loss unit in the Department of Psychiatry, and I find that a lot of overweight people overrespond to food cues. I teach them to relax in the presence of a food odor instead of getting frantic. I also teach people with lower back pain to relax. First I guide the person through head-to-toe muscle-loosening exercises. Then, when they're very, very relaxed, I give a hypnotic suggestion that I'm going to have them smell an odor, and the next time they smell it, they're going to be able to re-create this wonderful feeling of relaxation. I use a fruity scent with relatively pleasant or neutral associations, such as apricot, peach, or plum. I give people little bottles they can carry in purse or pocket."

If you can teach people to associate a scent with pleasant emotions, can you do the same with unpleasant ones? Certainly this happens spontaneously; for me the smell of Brut after-shave calls up whole worlds of adolescent agony. Using techniques similar to Schiffman's, couldn't convicts, for example, be taught to associate the smell of, say, steak with abject fear? Then guards in low-security, high-cholesterol prisons need only brandish a seared sirloin to subdue troublemakers. Or maybe things could be engineered so that all the guards would really need is a teensy spritz of Eau Sauvage on strategic pulse points.

Among those interested in the connection between scents and sensibility are commercial fragrance houses, the folks who formulate the scents for famous-name perfumes and household products.

Dr. Craig Warren is vice president and director of organoleptic—taste and smell—research at International Flavors and Fragrances (IFF), perhaps the nation's leading smellmeisters. "We have about 200,000 formulae in inventory," says Warren. "Everything from fresh butter to root beer to roses." A small but intriguing part of the business involves creating unusual aromas for clients: freshly baked bread, perhaps, or new car smell, which is mostly leather.

"At one point," Warren reminisces, "we had a call for a vomit odor, to be added to an experimental appetite suppressant. It was a technological success but a commercial failure." Even such near misses go into the fragrance vault, he says. (Although I can't imagine that this one will come up again. For what: Calvin Klein's Regurgitation? Upchuck of the Ritz? Puque by Lanvin?)

"From time to time we've been asked by museums to make a smell for a specific exhibit," Warren says. "For example, someone once wanted the odor of eighteenth-century London. We had to guess from historical evidence what the components might have been. I wasn't involved in the project, but I imagine it was a kind of coal-smoke-and-horse-manure smell.

"A typical request is for the smell of the ocean. That's impossible to do; the actual smell—seaweed, clams—isn't pleasant out of context."

Researchers in IFF's psychology program, Warren says, are studying how odors change mood and which odors evoke which moods. Such information would, he says, allow perfumers to fine-tune the mood created by a fragrance and better match advertising to the product. "For example, you wouldn't use a sensuous ad if the perfume doesn't test sensuous." But Warren is most interested in whether odor provides some sort of special sensual experience that can't be supplied by another sense.

"Odor is, after all, mediated by the area of the brain that also mediates sexual behavior, survival, and appetite; this should give odors some special attributes over other sensory modalities. We don't know for certain that pheromones [body chemicals that elicit an instinctive behavioral response in another member of the species] exist for human beings the way they do for animals. No one has isolated in human beings the odor of a bitch in heat." (I would definitely buy Calvin Klein's Bitch in Heat.)

When IFF chemists want to create a new smell—living rose, for example, one of their current projects—they follow the procedure Warren says is used by all such companies. The odor is pulled by air pump onto an absorbent material; it's deabsorbed in a gas chromatograph, an instrument that separates the components of odor. Then a mass spectrometer identifies the molecular structure of each component.

It was this procedure that enabled the company to hit on a newly voguish fragrance compound that, added to laundry detergents, makes clothes spun in a dryer smell as if they've been hung on a line. You know, analog drying. "The sun-dried cloth odor was done by hanging cotton cloth on a line, exposing it to the sun, extracting molecules from the cotton, determining their structures, and testing to see which evoked the sun-dried smell," Warren says.

Sociological trends affect fragrance popularity, says Avery Gilbert, director of Olfactory Science for Roure, Inc., a competitor of IFF. "Lemon-fresh has been in ascendancy for twenty years. And for the baby boomers, baby powder has been a powerful cultural artifact." Apparently they crave the smell long after infancy; several of the fine fragrances most popular among that generation—Canoe, Ambush, Shalimar—contain the same sweet note, Gilbert says. (Talk about the quest for youth! How about In Utero by Guerlain?) "One hypothesis I've heard is that the baby powder scent was so successful originally because it mimicked the smell of newborn babies themselves."

Empirical evidence suggests the connection between scent and memory is strong. How could I not invoke Proust's madeleine and lime-flower tea, the snack that launches seven volumes of *Remembrance of Things Past*: ". . . the smell and taste of things," he writes in *Swann's Way*, ". . . bear unfaltering, in the tiny and almost impalpable drop of their essence, the vast structure of recollection." I feel that way about a certain kind of weed-killer used on the lawns of my elementary school. One whiff brings back intense, idyllic childhood memories, even though Eau de Foliant is certainly not pleasant. In a recent experiment, Arnie Cann of the University of North Carolina found that fragrance forges a powerful link with memory, whether the smell is good or bad.

"I was interested in the possibility of using olfactory information as a way to help retrieve memory," Cann says. So he designed a two-part experiment. First, participants were briefly shown fifty photographs of people and asked to judge their attractiveness. At the same time they were subjected to either a nice floral fragrance or a stinky chemical—"hard to describe, but it made my colleagues unhappy," says Cann. Contrary to expectations, the pleasantness or unpleasantness of the odor had no effect on attractiveness ratings.

Two days later, the same people were shown 100 pictures and asked to identify the ones they'd seen before. "If the same odor was present as when they first saw the picture, recognition was high, whether the odor was pleasant or unpleasant," Cann says. "If there was no odor, or if we switched odors on them, the recognition wasn't as high."

"We've all had the experience of odors retrieving childhood memories," Cann says, "but nobody's shown until now that memory can be retrieved through systematic cues." The finding has practical applications, Cann says. "Perfumers could market a personalized scent; when people smelled it, it would create your own memory trace." (His could be called Cann Dew.) Some day, he speculates, these findings may be used to enhance courtroom testimony. "If in fact it could be determined that a distinctive odor was present at the scene of a crime, for example, and if you could reinstate that odor, the individual might be better able to recall the event," Cann says. "That's untested, but a natural extension of what we found." (Can't you just hear judges pounding their gavels and shouting, "Odor in the court!"? Of course, witnesses may be intimidated by threats of cement nasal spray.)

Other interesting applications may come from the work of psychologist Pete Badia at Bowling Green State University in Ohio. He's studying how well people smell—not how good they smell—in their sleep. "We know that the auditory system functions during sleep, but there's been virtually no work on olfaction," he notes. "Sleeping people don't always respond to the smell of smoke, so we weren't sure."

Volunteers who spent the night in the sleep lab were told that when they smelled an odor—generally jasmine or peppermint, delivered via a steady flow of air through a modified oxygen mask—they were to respond by awakening and pressing a switch taped to their hand. Sixteen percent pressed the switch on cue, although fewer than half of them took the trouble to wake up first. Even among the other 84 percent, subjects showed telltale changes in heart rate, respiration, muscle tension, and brain waves.

"Clearly the olfactory system functions during sleep, and clearly this has safety implications," Badia says. (Could I interest you in a Completely Organic Smoke-Activated Gorgonzola

Distant Early Warning System?) Next he's going to look at whether different fragrances have an effect on the quality of sleep at each of its five stages; he'll eventually attempt to manipulate restfulness and wakefulness. (You might want to get busy on a Strawberry Alarm Clock or a timepiece that releases some other scent—skunk, sweat sock—able to roust the heaviest of sleepers. "My God! It smells like 7:30 already! I'll be late for work!")

I'm working on my own mood-enhancing fragrance, designed to induce a devil-may-care attitude. I call it Eau de Hellwithit. And I thought Shimizu might be interested in a dramatic Japanese cologne called Eau Noh. Or maybe I should stop making scents.

Triumph of the Willies

□ ■ □

Fingernails on a blackboard. Hey, come back. I want to ask you something. Just seeing the words is enough to give you the willies and chase you away, right? But why, do you suppose? I think some science is going on here.

To learn more about the evolution of the deep frisson (I think we'd agree on the subjective symptoms: shivers, raised hackles, racing heart, gritted teeth), I took a haphazard but sincere nationwide poll that revealed an astonishing variety of triggers.

Fingernails on a—you know—chalked up the most votes. The next most often reported irritant was the whine of a dentist's drill. Others included the squeal of balloon skins, Styrofoam, Magic Markers, subways, or brakes; the bleat of a car alarm, the cracking of knuckles, and the hum created when a finger is run around the rim of a wineglass. One woman gets the willies from sibilance. "When I was a kid," she told me, "my father would torture me by chasing me around the house saying, 'succulent, succulent.' " (Are you thinking what I'm thinking? A Movie of the Week, with maybe Jane Seymour?)

People reported shivering at the feel of talcum powder between the toes, of angora or of rough pottery. Visual chillers include seeing someone pick fuzzballs off a sweater—a neurologist copped to that one—bite ice, or clip fingernails. Also mentioned as distressing were straws in soda cans and caterpillars ("because they look as if they contain too much liquid"). We're not talking phobia here; people weren't *afraid* of caterpillars or straws, just physically irritated by them. One guy said he can't bear the sight of Chinese pork buns "because they look too placental." (I nodded and smiled. "Humor him," I thought. "He may have a gun.")

Besides quaking over clawed slate, I myself get industrial-strength chills when I see people fold a piece of paper, then absentmindedly run the crease between thumb and forefinger. I'd rather commit hara-kiri than witness origami. I also share with at least two other Americans a horror of seeing anyone bite paper. (You think that never happens? Hah. In *The Big Easy* Dennis Quaid plays a police detective who destroys evidence by eating an envelope. Aaargh cubed! The only thing that kept me from running out of the theater screaming was the possibility that Quaid might take his shirt off again.)

I've always used the *willies* or the *chills*; others know this constellation of aversive responses as the shudders, the whim-whams, the heebie-jeebies, and (inexplicably) the juts. Anthropological folklorist Alan Dundes of the University of California at Berkeley says we're all wrong.

"I'm not sure there's a precise word for the blackboard-and-nails phenomenon," he says. "*Gooseflesh* probably comes the closest. The *chills* overlaps with disease; the *willies* is a scare." The latter term, Dundes says, may have come from a genre of macabre Victorian rhymes called Little Willies, after the main character in *Ruthless Rhymes for Heartless Homes,* an 1899 book by Harry Graham, a minor English poet. It contained such mad doggerel as

> Billy, in one of his nice new sashes
> Fell in the fire and was burnt to ashes;
> Now although the room grows chilly,
> I haven't the heart to poke poor Billy.

Billy, a.k.a. Willy, became the dead-baby joke of his day; "to give the willies" came to mean frightening someone.

Whatever you call it, the phenomenon seems universal. Anthropologist Paul Bohannan, former dean of social sciences at the University of Southern California, reports that while theirs is certainly not a blackboard jungle, the Tiv tribe of central Nigeria respond in the classic manner to the squeak generated when an unripe ear of corn is shucked.

Some argue that the response is strictly biological, the result of the way sound is processed in the inner ear and the part of the brain known as Area 44 (a great name for a disco). Says John Daugman, professor of psychology and of electrical, computer and systems engineering at Harvard, "The human auditory system is responsive from 20 to approximately 10,000 hertz—vibrations per second. That range is divided into about twenty-five frequency bands, each a third of an octave wide. The eighth cranial nerve, the auditory nerve, has about 30,000 nerve fibers. Each is selective to a different band; some overlap. If any one frequency is saturated with energy, the sound may become aversive." Elegant, but it doesn't explain fuzzballs and talcum powder.

"I get the willies when I see closed doors," Joseph Heller's novel *Something Happened* begins. "Something must have happened to me sometime."

Something must have happened, all right, way back at the start of this wacky cosmic chain letter we call human evolution. Such a universal response must once have had survival value. Indeed, every neurologist to whom I spoke (literally a handful) considers the willies to be a demure descendant of the primitive fight-or-flight response. With a rush of adrenaline, the autonomic nervous system mobilizes the body through local vasoconstriction and piloerection: capillaries in the skin close off; robbed of warm blood, the body pales and chills. Hair springs up on pedestals of goose bumps as tiny muscles around each follicle contract. (During a hairier epoch piloerection made us appear bigger and scarier—think of Don King—and perhaps provided better insulation against cold by trapping warm air near the skin. Dogs still respond to high-pitched sounds with raised hackles.) Our pupils dilate; our teeth are set on edge—perhaps a throwback to fang baring.

"It's a completely irrational response," says William McClure, director of the program in neural, informational, and behavioral sciences at the University of Southern California. "It's a 50,000-year-old set of emotions in conflict with a societal overlay of staggering sophistication. These stimuli appeal to ancient parts of our brain. A built-in mechanism says we should be concerned about this sound, but none of us remembers why anymore.

"Perhaps—and this is just conjecture—these sounds are similar to the rattle of a leopard's claws on a rock; the ancient message was 'get out and don't look back.' "

Psychologist Randolph Blake of Northwestern University may have come closest to unlocking the secret of the willies. He's the coauthor of one of the few articles written on the subject, a 1986 study in the journal *Perception and Psychophysics* called "The Psychoacoustics of a Chilling Sound."

Psychoacoustics is not the science of padded cells but the study of perceptual reactions to sound. "We broke down the sound of fingernails on a blackboard into its component frequencies," Blake explains. "It's a very complex wave form. Then we removed certain frequencies from recordings of the sound and had people judge how aversive these filtered versions sounded relative to the original. We found that it's not the high-frequency components that set our teeth on edge, as we'd thought. People rated the signal with high frequencies removed just as aversive as the original. When the middle-range frequencies were removed, however—2,000 to 4,000 hertz, where we're most sensitive—the sound lost its obnoxious quality."

Blake and his colleagues speculate that this is not an acquired aversion but a by-product of our biology. How did it get there? What naturally occurring events produce sound waves that mimic those produced by fingernails on a blackboard? The researchers compared several, and the wave forms most similar were those of primate warning cries. "Perhaps," Blake says, "we get the willies because the sound of fingernails on a blackboard mimics a biologically relevant warning cry that was part of the behavioral repertoire of our primate ancestors." A human scream also measures about 3,000 hertz.

Blake suggests that future williologists—not his word—might wish to explore how monkeys react to a recording of fingernails on a blackboard. Other interesting questions: At what age do children begin to find the fingernails-on-blackboard experience hideous? Are older people more susceptible to the willies than younger people are? Is one sex more sensitive than the other?

Blake reports that the recording never lost its power to make him shudder, even after he'd listened to it hundreds of times. Yet, he points out, chills aren't always unpleasant. "I get essentially the same autonomic nervous system response and find it pleasing when I'm listening to Tchaikovsky's *Pathétique*."

In a 1980 study, pharmacologist Avram Goldstein of Stanford found that about half of his 249 subjects regularly experienced a tingle or tremor, often accompanied by gooseflesh and throat lump, in the presence of such stimuli as music; moving scenes in a book, play, or film; beauty in nature and art; and heroism, observed or read about. These thrills most often began in the upper spine and back of the neck, spread upward over the scalp and face, outward over the shoulders and arms, and then down the spine. This pathway is perhaps determined,

Goldstein suggests, by electrical activity in the amygdala, a part of the brain involved in emotions and discharges of the autonomic nervous system.

To explore the role of endorphins, the body's natural opiates, Goldstein asked subjects to report the frequency, intensity, and duration of thrills engendered by a piece of music—first without drugs, then again after having been injected with naloxone, a drug that blocks endorphins. Naloxone significantly squelched thrills in three of ten subjects. Goldstein thinks higher doses would likely still the tingling spine in all subjects.

The hows may yield to probing, but not the whys. "It is curious," Goldman writes, "that evolution selected for and conserved this curious ability to feel a slight shudder when we are emotionally aroused." Pending further investigation, we're in the same boat as Trudy, the bag lady in Lily Tomlin's one-woman show, *The Search for Signs of Intelligent Life in the Universe,* by Jane Wagner. Trudy is visited by extraterrestrials who want to know about goose bumps. Where do they come from, and where can the visitors get some? Trudy can't answer the first question, but she takes the aliens to a play, where they are profoundly thrilled, enough to experience the desired piloerection. She's forgotten, however, to tell her space chums to watch the stage; they've gotten goose bumps watching the audience. "Yeah," says Trudy, "to see a group of strangers sitting together in the dark, laughing and crying about the same things . . . that just knocked 'em out."

A sweet view of our species. But disturbing questions about the willies remain. This thing with me and paper—what does it mean? Do I retain a dim, genetic memory of a time when huge tablets of twenty-pound bond roamed the savanna in vast reams, ruffling their pages ominously at my arboreal ancestors? Did Peking man stalk the prehistoric steppes with heaping platters of proto-pork buns and dim sum his enemies to death? I holler into the well of the past and hear only the echo of an obnoxious sound pitched at about 3,000 hertz. The rest is silence, as another Willy once wrote.

Can We Tock?

□ ■ □

I'm sick of hearing about my biological clock. I can't pick up a magazine or eavesdrop on a luncheon conversation without detecting its inexorable tick. No one, however, is telling me what I really want to know about this internal, infernal time-keeper. Does it come with a snooze alarm? What if it turns out to be a biological clock-radio: Is there any chance that my ovaries could suddenly begin blaring out traffic reports or "Born in the USA" while I'm at an important business meeting or arguing with my landlord?

Sure, I understand that basically we're all on Genetic Standard Time: the body's DNA programs precisely how long we remain fertile and indeed, barring accident, sets the hour of our bucket kicking. Science has recently found the masterclock of our daily rhythms, a section of the hypothalamus called the suprachiasmatic nucleus, not much bigger than a pinhead. It seems to govern the hundreds of bodily functions that ebb and flow in a regular pattern throughout the twenty-four-hour cycle of our waking and sleeping. These include body temperature (it peaks at 4 P.M.), levels of various hormones (high tide for testosterone is 9 A.M.), and tolerance of pain (greatest in the afternoon).

But how come we never hear anything about the other interior chronometers that are just as important as the Big Ben of babymaking and the Rolex of rhythms? What about, for example,

THE BOTANICAL CLOCK: Perhaps regulated by the pituitary, the Botanical Clock fixes the precise point in a relationship at which a man stops sending flowers. It is closely related to the ASTROPHYSICAL CLOCK, which controls the length of time a loving couple will have stars in their eyes.

THE TAUTOLOGICAL CLOCK influences redundancy, repetition, and redundancy. It is also responsible for making an otherwise pleasant friend or mate say things like, "Well, listen, you'll either get the job or you won't."

THE ONTOLOGICAL CLOCK: By maintaining a delicate balance of brain chemicals, this clock determines how long you remember the definitions of really rather simple words that won't stick in your head, like *nonplussed* and *ontology* (the one that does *not* recapitulate Phil Donahue; that is onto*geny*. Ontology is, of course, the branch of philosophy dealing with being, something I can tell you with great confidence, having just looked it up for the third time this year.) Nutritional deficiencies can cause some genetically vulnerable individuals to forget the difference between *oligarchy* and *plutocracy*.

Not all body clocks are governed internally. We all own a CULTURAL CLOCK, which has to do not with how soon we get fidgety at the ballet, but with how our culture views time. Never mind the globally acknowledged oscillations of the atomic clock, the transit of the sun, or even the vaunted suprachiasmatic nucleus: we pick up the tick tocked by our parents and grandparents. Thus, while in remote parts of Brazil it is acceptable to appear at a dinner party several decades late and feign surprise that the host and hostess have passed away, in some Swiss cantons it is considered rude for a guest to enter a room in such a way that his hat arrives before his shoes.

Intimately related to the CULTURAL CLOCK is the one that

controls differences in the way the sexes perceive time. I call it the GENDEROLOGICAL CLOCK, and who are you to stop me?

I must thank the late Lorne Greene for leading me to the discovery of this clock and, ultimately, to the Alpo theory of rapprochement between the sexes. One night I was watching his classic dogfood commercial, idly wondering whether those mutts get their residual checks figured in dog-years or people-years, when all of a sudden I thought, "Hey! I wonder if there are man-years and woman-years!" That could explain a lot. For example: A certain man tells a certain woman he will call "in a couple of days." Two weeks pass. He dials her number—and is shocked by her cold response. She's wrong to think he's been toying with her. He said two days—*and for him, that's all it's been!* Applying a formula very similar to the one Lorne Greene uses to figure out the age of a golden retriever, you can convert man-time into woman-time. (Roughly seven woman-days = one man-day, when you're dealing with relationships. Different formulae apply to work situations and personal growth issues.)

Looking back, I can see unheeded hints of these divergent perceptions of time in my own true life. When I was sixteen, for instance, my boyfriend and I stayed late at a riotous New Year's Eve party in a town about an hour from home. Rather than risk the roads that night, we stayed, quite chastely, with friends. I called my parents to tell them where I was and that I'd be back in the morning.

When we pulled up at 11 A.M. the next day, my father was raking so furiously in one spot that the lawn was bald. "Where the hell have you been?" he barked. "I *said* I'd be home in the morning," I whimpered. My father's reply, now famous in our family: "Eleven o'clock isn't morning where I come from!" Since I know for a fact that my father comes from Boston, I should have realized right then that man-time and woman-time are out of synch.

Maybe it's because so many little boys first learn about time from games like football, in which an announcement that there are ten minutes left in the final quarter means you have the leisure to catch several trout, finally finish *Sister Carrie,* jog a mile, shower, and reestablish yourself on the couch just as the two-minute warning sounds. Whatever the reason, it is a scientific fact that just because a man and a woman share a zip code, that doesn't necessarily mean they live in the same time zone.

The Novocaine Mutiny

□ ■ □

Gee, I can't imagine why people fear dentists. It must be the psychological tensions that arise when the subject touched upon is teeth, which, Freud explained, symbolize sexuality, power, and the best way to make a cheeseburger fit down your throat. It couldn't possibly be because dentists drill holes in your face, expose and twang delicate nerves, hum off-key, plug their excavations with evil-smelling goo, then make you bite blue paper. Or that deadening the pain created by the aforementioned activities involves plunging two inches of needle-sharp needle into your gums and freezing your mouth into a hideous rictus so that you worry about drooling in public for the next five hours. Some folks are just skittish, I guess.

I was especially spooked by a recent adventure in oral surgery that made a root canal look like a day in the country, even if the country is Lebanon. I don't want to talk about it. Okay, I'll tell you a few basic facts. All right, I'm installing an 800 number for those who want to hear every ghastly detail. Briefly, I underwent an apicoectomy (from the latin *apex,* meaning tip), a procedure in which an endodontist (from the Latin *endodontare,* meaning to hurt someone beyond the limits of human endurance and charge them $575 for the experience) slices open the gums, removes the tip of a tooth's root and any surrounding rot, fills the hole with cement (leaving his footprints and signature), sews it shut, and sends you home with inadequate painkillers. Then you lie around and swell. First I looked like Little Lulu, then Meatloaf, then Jabba the Hut. Sympathetic friends faked coughing fits to hide their amusement. My nose, cheek, and jaw grew so woefully bloated, in fact, that the straining sutures shredded my tender tissues into a bloody slaw. I don't want to talk about it. And when I called to say I was in pain, the receptionist said, "None of our other patients have complained about pain after three days." Uh-huh. "Grand Inquisitor's office. What? Well, none of our other heretics have complained about *their* autos-da-fé."

Since then, I've been a little gum shy about dentists. So I was especially pleased to talk with the people at the University of Washington Dental Fears Clinic. Since 1982, its team of two psychologists, four dentists, and two hygienists has been using a variety of behavioral techniques to help people with dental phobias so severe that their lives are disrupted. (According to Louis Fiset, D.D.S., who works at the clinic and is a research assistant professor at the UW School of Dentistry, one patient had to make four visits before she could get from the waiting room to the office. Another, blaming the disintegration of his marriage on the disintegration of his teeth, said he'd kill himself if he couldn't conquer the phobia.)

The American Dental Association estimates that half the people in the United States have some fear of dentistry, and the other half don't have the sense God gave a goose. About thirty million of us are phobic—so fearful that we shun dental treatment altogether. (According to Philip Weinstein, Ph.D., a psychologist with the clinic, extensive research has shown that the rate of dental phobia in Japan is double ours. That could be because, as dento-economists have known for centuries, a nation's dental fear index is inversely proportional to its budget deficit. Or it could have something to do with the traditional Japanese concern for "saving face," which may entail more extensive oral maintenance work.)

The first step for patients at the clinic is a session with a psychologist to determine the shape, scope, and genesis of the terror. Not surprisingly, most patients—whose average age is thirty-eight, though adolescents, children, and the elderly are also treated—say their dental fear developed in childhood.

"In 1986, we did a study of dental fear in Seattle adults," Weinstein reports. "Two-thirds of the people who said they were afraid became afraid as children." If you and your teeth grew up in the sixties or earlier, Weinstein says, you survived an era when dentists often drilled on kids without anesthesia; the common wisdom held that a child's nerves weren't fully formed, so the patient felt no pain. "We no longer believe those things, of course," Weinstein says. "There's been lots of new information about what pain is and how to control it in the last decade and a half."

Your emotional state, for example, can affect the physical transmission of pain. "Suppose I poke you in the knee," Fiset says.

"I'd run shrieking out of the office," says I, "since I'd be especially wary of a dentist who pokes me in the knee."

"This is a hypothetical, nondental situation," says Fiset, whose patience and good nature seem to be of Mr. Rogersian proportions. "If I poked you, you'd probably call that pressure, not pain. I stimulated your nerve endings, but the stimulation wasn't strong enough to cross the synaptic gap [the miniscule space between nerves bridged by neurotransmitting chemicals] and activate your pain receptors.

"But if you were terrified about what was going to happen next, your body's fight-or-flight response would kick in." My brain would signal my adrenal glands to release adrenaline, a hormone that also serves as a neurotransmitter. When the adrenaline bathed the synaptic gap in my knee, the poke that would ordinarily register as mere pressure would jump the gap and activate pain receptors. "When a patient reports pain and the dentist thinks there shouldn't be any, it could be because the person's upset." Or because her knee hurts.

Or it could be, Fiset says, because local anesthesia is inadequate—something that happens one out of every six times it's used, according to two recent studies done by clinic researchers. (Personally, I prefer express anesthesia to local.) "If a dentist thinks that because he put the needle in a certain spot he's achieved anesthesia," Fiset says, "he hasn't been properly trained. From patient to patient there's lots of variation in the location of nerves. To make certain you're numb, the dentist has to do a sequence of injections."

Once you've experienced pain in a dentist's office, you're conditioned in a Pavlovian way, says Fiset. "You begin getting upset as soon as you enter the dental environment." (All I have to do is see a copy of *Highlights for Children* and my teeth are on edge.)

At the clinic, patients learn coping responses to counter their physical and emotional upheaval. "We teach them how the mind influences the body—by secreting the chemical that's causing their heart to race, for example—and we teach them how to control the body through relaxation exercises, breathing exercises, and dissociation through guided imagery. We try to desensitize them to situations that evoke fear."

At the initial session, patients are asked to figure out their fear hierarchy. "Calling the office to make an appointment might be the first rung on the fear ladder," Weinstein explains. "Driving on the freeway to get there may be the second, followed by sitting in the lobby, then being told that it's your turn—escalating events that raise the anxiety level, all the way to drilling. Then we slowly expose them to what they're afraid of, and they practice the calming responses we've taught them. If a patient becomes too anxious to go on, we stop, go back to the beginning, and start over again, working up to the acceptance of dental treatment. This takes an average of three visits, but sometimes it takes eight or nine. In some cases we use nitrous oxide, antianxiety drugs, even intravenous sedatives. But drugs don't change behavior, so we still use the behavioral techniques."

Dental phobics fall—or are gently placed—into four categories, Weinstein says. Some are apprehensive about a specific procedure—drilling, say, or an injection. "I've worked with people who can't tolerate a syringe in the same room, even in a locked drawer," Weinstein notes. "Once I was working with a patient who noticed I had a bandage on my finger; when I explained that I'd had stitches, the patient said, 'You had to have an injection to get stitches'—and just that thought was upsetting."

(I was about eight when our family dentist first sprang the news that if I didn't want my tooth to hurt when he drilled, he could—and these were his exact words—"Jack Frost it up for me." Then he revealed that Jack was hiding in a syringe taller than I was. I didn't see how such a needle could possibly fit in my mouth without piercing the back of my head and pinning me to the chair like a dead butterfly, so I politely refused. I was in college before I submitted to a shot of novocaine. Let's keep that just between us, okay?)

The patient afraid of shots—let's call him or her "he"—is given a detailed description of the procedure and then allowed to rehearse it step by step. He is in control; he can stop at any time and start over. If he becomes anxious, he can apply relaxation and breathing techniques. At each increment, the patient rates his anxiety on a scale from 1 to 10, with 10 equaling "I'm going to run out of the room." The dentist repeats each step until the anxiety level drops to 3. Step one: The patient holds the syringe in his hands with the cap on so he doesn't see the needle. When he's ready, he moves on to holding it with the cap off and the needle exposed. (See that little space? That's where my anxiety level reached 10 and I ran out of the room.)

Then the dentist places the capped syringe up against the patient's gum and leaves it there for ten or fifteen seconds, a length of time negotiated beforehand. If the patient is ready, the dentist tries the same thing with the needle cap off. Finally, the dentist makes a deal with the patient: He'll put the needle up against the gum and count to ten, and if the patient hasn't given the signal to go ahead with the injection, the dentist will remove the needle and rehearse again. "Usually, patients allow us to give the injection," Fiset says. "But they know we'll never begin the procedure without their permission. This process usually takes less than an hour."

Another group of dental phobics experiences a generalized anxiety so severe that they can't sleep or concentrate and may behave irrationally; they feel embarrassed and childlike. A small percentage of patients are convinced that their bodies can't tolerate dental treatment—they're afraid they won't be able to breathe or that they'll have a heart attack and die. Both these groups respond well to information and practice sessions. For people with fears about their bodies, researchers are looking into the use of beta blockers, heart medication also used for extreme stage fright.

Most often, however, the key issues for phobics are trust and control, Weinstein says. "Most of our patients report that they've had bad experiences—the dentist wouldn't stop when they needed a rest break, or he insisted they couldn't possibly be in pain because he'd given them enough novocaine—that made them feel trapped, vulnerable, powerless. We allow them to have that control."

That's what Vince Santo Pietro wanted. Before March of 1988, Santo Pietro, thirty-two, manager of visitor education at the Seattle Science Center, a hands-on museum, hadn't been to a dentist since he was seventeen. "I was having my first filling. The dentist promised me it wouldn't hurt. But it hurt a lot—an indescribable pain. He wouldn't believe me; he said

I was just afraid and kept drilling. Then and there I vowed I would never go back."

He stuck to his vow, though he felt uncomfortable. "I have a masters in chemistry and I develop science programs—I've been taught to think logically, and this wasn't logical. I kept it a secret. It was a constant burden on my mind, compounded by the fact that I had a wisdom tooth I knew needed to come out. But every time I considered going to a dentist I thought no, I'll probably need a filling and it will hurt. I was convinced a dentist wouldn't do what I wanted, that he'd insist on drilling before he'd fix the wisdom tooth. A friend who's a psychology major at the University of Washington told me about the clinic and sent me a brochure. I kept it for six months. Finally, I couldn't deal with the pain from the wisdom tooth. I decided it was time to deal with the fear.

"The first visit was great. I could tell that Tracy Getz, my psychologist, believed me and didn't feel my fear was stupid and childish, the way everyone else said. I relaxed. But I was still skeptical.

"I came back for an orientation to the dentist's chair. Tracy stayed with me. Dr. Fiset was wonderful. I just sat in the chair and he talked about his instruments and didn't do any work. I was surprised that the office didn't look like my old dentist's—no drill run by big rubber bands, no spittoon. Dr. Fiset told me that the dentist I'd seen in the seventies had been using vintage fifties equipment. That made me feel a little better.

"They told me I'd have control over what went on in my mouth, and that was important. Another thing that was important is they let me ask questions, always waiting to find out if I needed more information."

Santo Pietro made it through a cleaning; the hygienist did one part of a tooth at a time, let him stop whenever he liked, and gave him a mirror so he could watch the procedure. "And Tracy helped me relax, reminding me to keep breathing. I'd close my eyes and visualize pleasant things—a nice weekend I'd just had, for example—to remove myself from the fear.

"Then I cracked a tooth and did need a filling and reconstruction. It took several visits to work up to it. I had an orientation, where the dentist showed me exactly how he'd go about it. He had to prove to me I wouldn't feel a thing before I'd let him work on me. My old dentist would inject me with one cartridge of novocaine, and that was it. But Dr. Fiset was able to measure the exact amount of anesthesia he'd need. He had an instrument that applied an electric current to my gum. He'd turn it up gradually, and I'd tell him when I felt a sort of tickle. Then he gave me anesthesia and repeated the same process. After two minutes I couldn't feel the current. He'd figured out how long he needed to wait until I was numb, and how many cartridges it took. The second visit was a rehearsal—I had anesthesia but no tooth work. The third time, he actually started drilling. He wanted to drill for the count of five and then stop, drill for the count of five and stop, and so on. I wanted to start with the count of three. Eventually I worked up to the count of five. I controlled the length of the pauses. Once I asked to rest for a while. There was no pressure; he just went and did something else.

"When I went back for the wisdom tooth extraction, we went through the same procedure—testing how much anesthesia I'd need, rehearsing, pausing now and then. After it was over and I realized I'd had everything done that I'd needed, I felt great. I hadn't felt 100 percent whole because I hadn't dealt with the dental fear. Facing it made me want to deal with other things in my life that bother me—and made me better able to help others with their fears." Santo Pietro loves being able to tell kids who visit the museum that his tooth was reconstructed with the same kind of cement used to glue the tiles on the space shuttle. "The joke is that it's going to fall out, right? But NASA's fixed that problem."

For the mildly to moderately afraid, distractions may help with dental phobia. Music isn't effective, Weinstein says, unless you haven't heard it before. He suggests patients bring comedy tapes to the office. (I wonder if the truly petrified should try watching movies—say, *Marathon Man,* in which Laurence Olivier as a Nazi dentist drills Dustin Hoffman sans anesthesia.) Laughing gas is used as a distraction less often than you might expect.

"The main point of nitrous oxide is to make patients less anxious," says Fiset. "The mind tends to drift, you lose track of time, you don't attend to the environment that's causing anxiety. It gives some people a sense of well-being, and at some concentration levels, giddiness. But not all patients do well under nitrous oxide. For some, there are paradoxical effects; they become agitated, frightened, and out of control. We use it on a minority of patients. Some experience no effects at all, or aren't relaxed until the gas is administered at very high doses. But at high doses, you can run into dangers—nausea, vomiting, dysphoria, and disinhibition.

"Few controlled studies have tested the efficacy of nitrous oxide, though there's lots of folklore and poorly designed research," Fiset says. "We've designed a study that removes the possibility of clinician bias."

I know what you're thinking: that the researchers plan to use themselves as subjects, inhaling deeply and recording their own responses, such as "Hey, I'm not anxious at all!" and "Pass the Twinkies" and "Okay: A rabbi, a priest, and an orthodontist are in a lifeboat. . . ." Don't be silly. The three-year study, funded by the National Institute of Dental Research, will involve sixty volunteer patients who visit a community clinic to have teeth pulled. "We have an interesting machine that allows us to administer either nitrous oxide or room air at the flip of a switch," Fiset explains. "It's a double-blind study; the dentist administering the gas won't know which the patient is getting until after the patient responds." Researchers will measure patients' heart and respiratory rates during the extractions, and videotapes of the patients will be analyzed for differences in behavior under the two conditions. Once it's known whether the patient received laughing gas or air, he or she will get the opposite the next time as a control.

I myself can't get enough laughing gas, though a moderate disinhibition is the price of giddiness. During the disaster mentioned above, for example, the endodontist tapped on my tooth. "Come in!" I warbled wittily through a mouthful of his hand. He didn't even crack a smile (except mine, of course). Absolutely no response, which only made me giggle in embarrassment. A few more snorts of gas later I gestured for him to halt the procedure so I could ask him if he'd ever heard the poem "Do not go dental into that good night" by Drillin' Thomas. He rolled his eyes, sighed, and resumed carving my gums. I was totally mortified. But soon—I couldn't help it—I attempted to sing "She'll be gummin' 'round the mountain." The doctor remained unmoved, but I was in stitches, as it were. (This is drugs. This is your sense of humor on drugs. Any questions?)

Finally, I would like to state for the record that none of the negative comments in this column apply to my regular dentist, the understanding and personable John Siegal, D.D.S., a very hot number (when he says you're numb, you're numb) who always gives me a free toothbrush in a designer color, never sings along with his excellently chosen tapes, and should have no cause to pull a Laurence Olivier on me when he reads this.

From Hair to Eternity

□ ■ □

A forty-year-old woman showed up at the Indiana University dermatology clinic complaining of a terrible coif.

Her waist-length hair had swiftly, inexplicably, become horribly snarled, swirling upward into a dark maelstrom, a seaweedy mass. The less charitable among you might have called her Kelp Head.

Dermatologists Julia Marshall and Colleen Parker recognized an extremely rare syndrome called felting—the sudden, massive tangling of scalp hair, first documented in 1884 by a British doctor named LePage, who called the condition *plica* (from the Latin for "fold") *neuropathica.*

I'm sure that when you first hear about a disease, you instantly think the same thing I do: (a) How long do I have to live? and (b) Who should host the telethon? With these questions in mind, let's obsess together not only about the venerable plica neuropathica, but also about some newly discovered medical problems. As if the waning days of the twentieth century weren't malady-intensive enough, I'd like you to meet Seatbelt Gland, Toothbrush Throat, and Microwave Ear.

But first, back to the woman with nervous hair—let's call her Knotty Marietta. I sense that you're worried felting might strike you in public—say, in the middle of delicate financial negotiations. Impeccably groomed, you turn, midsentence, into Elsa Lanchester. No. Felting occurs in the privacy of the bath. Marietta had simply washed her hair as usual when, for the first time ever, that wacky batch of protein pulled a Medusa.

The medical literature dealing with hair felting is meager, Marshall says. (Except, of course, for the eighteen-volume *Severely Matted Hair That Changed the World.*) The first reported patient was a seventeen-year-old girl whose hair had suddenly become a hard, hideous roil after she washed it. LePage decided that because the girl had what he called "hysterical tendencies," her problem was "high nervous tension that found vent in the hair itself." Nine out of ten cases reported from the Victorian era to the mid-1950s were attributed to "nervous forces" or "hysterical tendencies." (My guess is that the patients became hysterical *after* they stepped out of the shower looking like Wookies.)

"Nervousness has nothing to do with it," Marshall says. "The problem is mechanical. From root to end, the cuticle, the top layer of the hair, is made up of cells that line up one on top of the other like shingles on a roof. As hair gets long and old, each cell can flip up the way an old shingle does. That makes cells more susceptible to injury caused by the mechanical action of shampooing. With vigorous scrubbing, the hairs, each with different lifted shingles, interdigitate." Marshall points out that this is the way felt fabric is made: wool fibers are soaked and agitated until they're tightly matted. Hence the syndrome's name, coined in 1970.

Because her hair was so long, Marietta had had plenty of time to flip her shingles. Then came the shampoo that broke the camel's back. "It's likely," says Marshall, "that her cuticular irregularities interdigitated." She calls the diagnosis of hysteria "silly and chauvinistic." (All reported cases of felting are in women, chiefly, Marshall thinks, because few men have hair long enough to allow premarital intercuticular interdigitation. If beards grew long enough, they could felt, too, she says; the

Smith Brothers are darn lucky they didn't end up looking like a mangrove swamp.)

Marshall proposed a shearing. "Usually the dermatologist's approach is 'when in doubt, cut it out.' " But Marietta's mane was twenty-four years in the making; the thought of a haircut distressed her. She chose, Marshall says, to attempt manual separation. "For two hours a night, she and her son lubricated the hair mass with olive oil and separated it with knitting needles." After two and a half months of picking her locks, Marietta appeared before Marshall with clinically normal hair. (I've seen fig. 1 and fig. 2. Now, her dark hair resembles Rapunzel's. When she first came to the clinic, it looked like a cross between a tornado and Bigfoot.)

Marietta blamed a particular shampoo, but felting can happen with any brand, Marshall says. "Some variables that affect felting include chemical treatments like bleaching and perming, fine or dense hair, and heavy scrubbing. If you have long hair, it's better to shampoo it in sections instead of piling it on top of your head. Cream rinse decreased felting in the laboratory." So shower in a laboratory whenever possible.

Just because only fifteen cases of felting have been officially reported, that doesn't mean it can't happen to us. If he were alive, Larry Stooge would make a good celebrity fund-raiser; among the living, Crystal Gayle is probably the star with the most at stake.

In the case of global felting, the hair's apparent. The trouble spotted by Michael Bornemann, chief endocrinologist at Tripler Army Medical Center in Honolulu, is much subtler. A patient came in with flulike symptoms and neck soreness that suggested an inflammation of the thyroid gland. "But a thyroid scan showed abnormalities on one side only," Bornemann recalls. "In thyroiditis, usually caused by an infectious agent like a virus, both lobes are affected."

Under questioning (truncheons were unnecessary in this

case), the patient revealed that the harness-type seatbelts in the used car he'd bought two months before hit him on one side of his neck. "There have only been a couple of cases of physical injury causing thyroiditis," Bornemann says. "One was a case of 'martial arts thyroiditis' reported by a physician who'd taken too many karate chops to the neck." Bornemann could only conclude that his patient's thyroiditis was caused by the rubbing seatbelt. Anti-inflammatory drugs worked swiftly. (Mercifully, they didn't have to shoot the car; a minor adjustment of the harness solved the problem.) Bornemann warns that owners of older cars are the most likely candidates for Seatbelt Gland; the belts in newer cars are more sensibly positioned and easily adjusted.

Bornemann was surprised not to find any other mention of such problems. "There's an extensive literature of seatbelt-related injuries," he says. (Of course, most of us are only familiar with Hemingway's *The Old Man and the Seatbelt.*) According to Bornemann, many of the problems are abdominal. I'd have guessed that the majority are back injuries sustained by parents on long car trips through desolate areas who twist too vigorously as they snarl into the backseat, "I'm thirsty, too. What do you want me to do, stop and milk a weed?"

My nominee for Seatbelt Gland telethon host is Geraldo Rivera because he, too, is a pain in the neck. If he's unavailable, then David Hasselhof, Jerry Van Dyke, Dean Jones, or any other actor who has played opposite an automobile would be appropriate.

I don't know how many times I've mentioned this, but apparently the message just isn't getting through: *Do not try to induce vomiting or scratch an itch by shoving a toothbrush down your throat, because you may swallow it.* I say this out of love, believe me.

Surgeon Allan Kirk of Duke University will back me up. Within the past five years, he says, the Duke Medical Center has seen several cases of toothbrush swallowing. Two patients were drunk and trying to rally a therapeutic spew. Another suffered a violent coughing fit while brushing his teeth, and thank God he wasn't cleaning the toilet. In all three cases, the toothbrush lodged in the patient's stomach. Kirk and colleagues rounded up the strayed tools of oral hygiene using an endoscope, a tube inserted down the throat.

"Interesting things have made it through the gastrointestinal tract," says Kirk. "Dog chains, fourteen-inch-nails. Intuitively, though, we thought a toothbrush wouldn't. And a check of the literature showed that in thirty cases of toothbrush swallowing reported, none has passed." Most ingestion of bizarre objects is part of severe mental illness, Kirk says, at least when performed by an adult. He read of a patient in the Soviet Union who swallowed sixteen toothbrushes and a spoon handle, but that was before glasnost.

Kirk hastens to point out that toothbrush swallowing is not an anomaly endemic to North Carolina. "The cases were clustered," he says, "not because they're epidemiologically related, but because being in a university area gives you a critical mass of people who are inexperienced and might try something odd." Leon Spinks might make an interesting spokesperson; he's certainly at risk for accidental toothbrush ingestion.

Dedicated though the doctors I've mentioned undoubtedly are, none, I should point out, felt moved to swallow foreign objects, pinch his thyroid, or stick her hair in a Cuisinart to learn more about the ills reported. James Lin, however, head of the bioengineering department at the University of Illinois at Chicago, was so interested in microwave auditory phenomena that he gave himself a dose.

"An intriguing thing was reported during World War II," Lin says. "When radar joined the war effort, people used to warm themselves in front of it. They said they could tell when the radar was turned on by the sound it made. But they shouldn't have been able to hear the radar's microwaves which, like light, are part of the electromagnetic spectrum."

Over the years, Lin says, the same phenomenon was reported by military personnel, employees in communication industries, and people living near large radar installations. They all heard a sort of low-pitched tone or humming in their head. Sure, it could have been their hair, revving up to felt. But Lin thought they were indeed "hearing" microwaves.

"I decided to test it on myself," he says. He ran an experiment in which a beam of microwaves was directed to the back of his head; he signaled to observers when he detected a hum. Sure enough, he could tell when the beam was on and when it was off.

Because he draws the line at probing his own brain with electrodes, Lin recruited small animals for the next part of his research. He confirmed that we do indeed hear microwaves, but not in the ordinary way.

"For the short period of time that brain tissue is subjected to microwaves," he explains, "it undergoes what I call thermoelastic expansion—the rapid absorption of energy converts to heat. The rise in temperature is miniscule—a millionth of a degree in a millionth of a second. Nevertheless, it launches a thermoelastic pressure wave that's conducted through bone or tissue to stimulate the inner ear."

The hum causes no problems at threshold levels, Lin says. "But at higher levels, as in the directed energy devices that were to be associated with the Star Wars project, brain damage will result."

Microwave auditory phenomenon is triggered by radar and other pulsed microwave detection and communication devices. Those of us who aren't in the military or who don't work in certain parts of an airport needn't worry—or so Lin says. "This isn't something your oven can cause." But I thought maybe you'd want to know so you can decide whether or not to wear a hard hat when you're nuking the morning flapjacks.

The thermoelastic wave theory suggests that a telethon for Microwave Auditory Phenomenon should be hosted by a hot-head. I suggest John McEnroe, with the Pillsbury Dough Boy in the Ed McMahon role.

Gag Reflex

□ ■ □

Laughing a hundred times a day provides a cardiovascular workout equivalent to ten minutes of rowing, according to William Fry, professor emeritus in psychiatry at Stanford University Medical School. And here's the beauty part: "You don't need equipment, you don't need a nice day, *you don't need humor,*" Fry says. (Italics mine.) (Well, okay, you can share them.)

Fry has been exploring the physiological effects of humor since 1953—and are his arms tired! But seriously, folks. He was studying the impact of honest mirth on blood pressure—laughter, he found, boosts cardiovascular fitness by lowering blood pressure and heart rate after briefly raising them—when he decided to experiment with simulated merriment. "I asked some of the subjects merely to pretend they were laughing at something funny, and I observed the same cardiovascular benefits, including exercise of the respiratory muscles, as in genuine laughter."

In light of this discovery, I frankly don't know why I should kill myself trying to give you a chuckle when faking it would do the trick. If you'd be so kind as to feign audible amusement at the end of each paragraph, you'd be doing us both a favor.

Fry was one of the first to discover some of the other, mostly salutary, bodily effects of busting a gut: it disrupts brain wave patterns, changes our breathing rhythms so that we expel more air than we take in, reduces pain perception, stimulates the production of hormones that increase blood flow, and strengthens the immune system by increasing the activity of cancer-fighting natural killer cells. And in 1989, Fry and colleagues published the results of a study showing that a "mirthful laughter experience" reduced levels of the hormones that help create stress.

Five healthy males watched a sixty-minute videotape of the comedian Gallagher. (If you aren't familiar with him, just wait patiently—he's bound to turn up on "Hollywood Squares.") Five others rested quietly, in a similar but Gallagher-free room on a different day. Blood was collected from each chap through an intravenous tube every ten minutes—for the merry men, three times before the videotape, six times in flagrante, and three times during a thirty-minute recovery period. (A conservative estimate for recovery from Gallagher, don't you think?)

The laughers showed lower blood serum levels of hormones associated with stress: cortisol, dopac (a chemical produced when dopamine is metabolized), epinephrine, and growth hormone, all of which can work to suppress the immune system. Laughter, it seems, can weaken or reverse the classic stress response.

(One important result of Fry's research promises to be the development of more objective criteria for film critics reviewing comedies. None of this vague "two thumbs up" stuff. Soon we can expect " 'I laughed until my pulmonary gas volume was markedly lowered!'—Gene Siskel" and " 'Less cortisol secretion than a barrel of monkeys!'—Roger Ebert.")

Fry has yet to ask people to *pretend* they're watching a Gallagher tape, but his work suggests that fake laughter may affect hormone secretion, just as it does cardiovascular activity.

Nobody knows yet exactly how laughter, or its counterfeit, does the job; it may, Fry says, encourage the production of endorphins, the body's natural opiates. And no one is entirely sure why this particular piece of behavior—responding to drolleries with the spasmodic expulsion of breath accompanied by short bursts of inarticulate braying—became part of the human repertoire in the first place.

"Humor has been a documented part of life for at least 5,000 years," Fry says. "Ancient Egyptian wall paintings depict slapstick and jokes." (One frieze-frame translates roughly: A papyrus merchant goes into the temple of healing and the priest says, "You're crazy." The merchant says, "By Ptah, I want a second opinion." The priest says, "You're ugly, too.")

But scientists haven't pinpointed precisely why laughter was evolutionarily advantageous. Was it because, as Freud believed, laughter is the healthiest way to exorcise repressed thoughts and feelings? Or because those prehistoric humans who spit rather than laughed to show they appreciated *le beau jest* soon died out (or were murdered)? Perhaps females were more apt to mate with those males displaying the least moist response to humor. (A dry wit may have been appreciated even by protohominids.) An important principle among our early ancestors may have been "Bleep 'em if they *can* take a joke."

Most likely, laughter emerged as part of a threat display—bared teeth and a scream—by dominant primates and protohominids. "The fact that gorillas and chimpanzees perform a sort of silent laughter suggests the early origins of mirth on the evolutionary tree," Fry says. Since laughter appears to have its roots in hostility, a more apt epigram than "Man is the only animal that laughs" might be "Man is the only animal that heckles."

Fry notes that it's hard to get grants to research the effects of laughter, even though the work is obviously vital to the relatively new field of psychoneuroimmunology, the study of how the mind helps to heal the body. In fact, Fry says, the only government in the world with an official humor section, something like the National Institute of Mental Health, is—you guessed it—Bulgaria, which even has a museum devoted to the subject. The House of Humor and Satire, in the town of Gabrovo, is heavy on the political cartoon. (I'm checking with its curator, Dr. Boris Tateerz, to see if the museum houses the earliest coal-powered joy buzzer.)

If you consider engaging in bouts of mock laughter beneath your dignity, you'll be pleased to learn that a phony smile may be good for you, as well. Several new studies show that facial expression—real or simulated—can change a person's mood. Smiling seems to make people happier and frowning seems to make them sadder, even if they don't know they're performing either act. In a recent study at the University of Illinois, student volunteers who thought they were testing theories of psychomotor coordination were instructed to clench a pen between their teeth in such a way that they created either a smile or a pout. Subjects felt happier and rated a series of cartoons funnier when they were in (unwitting) smile mode than they did when they were (unconsciously) pouting.

Charles Darwin was the first to suggest, in his 1872 book *The Expression of the Emotions in Man and Animals,* that facial expressions don't just reflect internal states but help cause them. (Doesn't Darwin seem to be the first to suggest everything? "Hey guys! Don't you think that allowing free play of the appropriate facial expression deepens an emotion, while suppressing the facial expression mutes the emotion? Hey guys! Don't you think species evolve over time through a process of natural selection? Hey guys! Don't you think fish and chips would probably go together better than this fish and parsnips you've got here??") But nineteenth-century Science rolled its eyes and twirled its forefinger in front of its ear.

Darwin's been vindicated, though, by the research of people like Face Czar Paul Ekman of the University of California at San Francisco, perhaps the world's leading researcher on the subject. In 1984 he was the first to show that when people mimic the facial expression associated with happiness, fear, or disgust, without the feeling, they experience the same bodily changes—in heart rate, skin temperature, muscular tension, and brain waves—that the genuine emotion creates.

Psychologist Robert Zajonc (pronounced ZI-ence) of the University of Michigan has suggested a reason that a fake smile has the same uplifting effect as a real one. Both involve contracting precisely the same forty-two facial muscles, an act that constricts veins in the face, decreasing blood flow to the sinus cavity and thus cooling it. The high volume of blood flowing toward the brain through the sinus cavity is also cooled, and so, ultimately, is the brain itself. Zajonc theorizes that this temperature drop affects the hypothalamus, the portion of the brain that regulates body heat (and perhaps other steamy *films noirs*) as well as emotions. In related experiments, Zajonc has demonstrated that lowering the temperature of the brain seems to elicit pleasant feelings, while raising the temperature seems to create unpleasant ones, perhaps, he conjectures, because a hot brain produces more of the chemical serotonin, excesses of which are associated with depression. (The reason Siberians aren't clicking their heels in perpetual glee while Tahitians sob nonstop, Zajonc says, is that the body automatically compensates for weather, as opposed to internal, changes.)

All this time we've been letting a smile be our umbrella when we should have let it be our thermostat! I wouldn't mind living in a world full of people making a mood-altering face instead of taking a mood-altering drug. Shoot, I'd walk around biting a Bic if it would give me a lift. As Shakespeare might have written after learning of the Illinois study:

> When in disgrace with Fortune and men's eyes,
> I all alone beweepe my out-cast state,
> I shove a quill betwixt my pair'd incis-
> Ors, and suddenly, you know, I feel just great!

Some of you may welcome the laughter-without-humor option. You may agree with Aristotle, who said, "People take more pleasure than they ought in amusement and jesting." Or was that Nipsy Russell? In any case, you're probably ready to bag the rowing machine and get down to the serious business of a 100-rep Mirthless Laughter Experience. If you haven't exercised in a while, don't guffaw without first checking with your physician. For a few days, warm up with three minutes of giggles and level off at a chortle; you'll be ready to split your sides in no time. Remember that a titter is especially good for those pesky pecs.

VEGETABLE

□ ■ □ ■ □ ■ □ ■ □ ■ □ ■ □ ■ □

Some scientists believe that primitive pine trees brought about the extinction of the dinosaurs, who ate them and died of constipation. The moral: Never underestimate the power of produce.

□ ■ □

The Ballad of the Green Bean Beret

□ ■ □

The Marines were looking for a few good shrubs. So they beat the bushes until they found one that beat back.

The Rambo of the vegetable kingdom is a plant called trifoliate orange, marketed as Living Fence by Barrier Concepts of Oak Ridge, Tennessee, and nicknamed P.T., for Pain and Terror. A fast-growing member of the usually mild-mannered citrus family, P.T. looks simply dense, not dangerous. (Although it does pack a pistil in the white blossoms that grace its shiny leaves and give way to green, golf ball–sized fruit.) What appeals to the Marine Corps and other Barrier Concepts customers like the CIA and NASA are P.T.'s concealed weapons: four-inch stiletto thorns that make the feisty foliage at least as discouraging as any nonliving fence for a fraction of the price—about $3 to $5 a foot for P.T. versus about $42 a foot for chain-link. And when mature, the hostile hedge is so thick it can stop a jeep.

P.T. is part of a growing branch of security concerned with growing branches. Though most of the breakthroughs in the world of barriers and surveillance are still electronic exotica, vegetation is making a strong showing. There's even a new discipline called security landscaping, a form of antiterrorist gardening that recruits flora and topographical features into Nature's own secret service.

"Shrubbery's in," says Bob McCrie, assistant professor of security management at John Jay College of Criminal Justice in New York and, for nearly twenty years, editor of the bi-weekly *Security Letter.* "The old is new again." P.T., which he employs as a visual aid in his introductory class, Security 101, was used extensively in the South a century ago to pen live-stock.

In Jamaica, McCrie notes, farmers have been using a form of natural security—the thyme alarm—for generations. "They plant a kind of thyme called seven-scent mint around their property," says McCrie. "When an intruder steps on it, the farmer is alerted by the intense odor that's liberated." (And of course you remember the line in *Hamlet* spoken by the screwy, herb-strewing Ophelia: "There's rosemary, that's for reconnais-sance.") In Indonesia, Guyana, parts of Brazil, and several other countries, McCrie says, dried cornstalks have long been used as burglar alarms; they crackle if anyone steps on them.

P.T., which has no audio, is rooted in the rural youth of Richard Shepherd, one of three men who founded Barrier Concepts four years ago; now, at sixty-eight, he's the com-pany's vice president of nursery operations. "We were watch-ing the news on TV, and we saw them backing a truck across the White House lawn as a security barrier. And we thought, that's ridiculous with all the technology we have." Shepherd and colleagues William Crisp and Jack Elkins came up with several dandy devices, like their patented high-security vehicle barrier: a pop-up steel plate, concealed in the roadbed and activated by electronic sensors, able to stop a 15,000-pound truck moving at 50 miles an hour. Then Shepherd, a Tennessee native and World War II veteran, remembered a particularly impenetrable thorny thicket of his boyhood and the spiky hedgerows that hampered Allied tanks in the Normandy inva-sion. He proposed such a no-tech barrier.

"It took me a year of combing the backwoods to find the

plant," he says. "It was practically extinct. It's been growing in Tennessee for a couple of hundred years. We think maybe it was brought in by Chinese building the railroads. We took some seeds and started from scratch to find out what makes them grow—I can't reveal our methods—and what kills them; that's cow manure, because it has too much nitrogen."

Today business is flourishing, though to what tune the privately held company won't reveal. Half a million plants grow in the main nursery, according to marketing director Jim Passmore, and three more nurseries in California, Texas, and Florida have subcontracted to raise more fence. Customers usually buy 36-inch plants, he says. Set 18 to 24 inches apart, they'll form a hedge 3 feet deep and 6 feet tall in a year or so. Unless new growth is trimmed, P.T. will reach a maximum height of 25 feet in temperate climates.

The topiary terror is catching on. "The Marines," says Passmore, "our biggest customers, are using it in conjunction with chain-link fence around ammunition storage areas, airfields, and guard dog compounds, to keep people away from the fence so they won't aggravate the dogs. The CIA put it on both sides of the entrance to their facilities; it's intended to channel protesters and demonstrators into the street so they can be controlled." NASA uses P.T. to keep nosy deer away from tracking stations; chain-link fences apparently interfere with telemetry. And since salt air is eating the chain-link at Cape Canaveral, NASA's thinking of surrounding the place with P.T.

Steve Keller, a Deltona, Florida, security consultant who specializes in protecting historical sites and art collections—the Vatican is a client—recently recommended to a prestigious museum that P.T. be planted near a constantly vandalized wall. "I have to put nature to work," he says, "because I can't put razor wire around a museum or public building."

Keller practices the art of security landscaping; he recently delivered a well-attended lecture on the subject at the annual convention of the American Society of Industrial Security. "Landscape can be used to conceal electronic devices like motion detectors," he explains. "It can create physical and psychological barriers—a hedgerow of thorny bushes screams out, 'Don't go in here!' and gives more credibility to the security person who challenges an intruder. It can create a no-man's-land where a person is slowed down by a first line of barriers, thrust into an area of high visibility, then slowed down again by another barrier as he approaches the protected space."

Security landscaping is a hot subspecialty for two reasons, Keller says. The first is the threat of terrorism. "It's drastically changed the landscape in Washington, D.C., for example. There's an Indiana engineering company, Everett I. Brown, with a computer program that allows them to look at landscape characteristics around a building and at the composition of the building—the thickness of the walls and the size of the windows—to determine what effect a bomb blast would have. Then they sculpt the landscape to deflect or absorb a blast." The second reason is a somewhat subtler form of terrorism—killer litigation. "The threat of lawsuits is changing the shape of shopping malls and parking lots and corporate headquarters. Property owners are being held accountable for failing to foresee crimes which could occur on their property."

The natural security movement doesn't depend solely on ground troops; there's also an air force. "Birds are big," says McCrie. U.S. Army bases in Germany use geese as watchbirds, because they make such a ruckus. "They're slow to start but hard to turn off—and they make good meals when their service in the field of protection comes to an end. In Turkey they prefer pheasants, not only at military bases, but at medium-security prisons."

Goose use has a long and noble history. In 387 B.C., a sneak attack on the Romans by the Gauls was allegedly foiled when a flock of sacred geese raised hell and roused the garrison. Thereafter the Romans annually paraded a commemorative golden goose to the capitol; presumably the vigilant flock was awarded the Croix de Foie Gras. You know those Romans—anything for a gaggle.

McCrie also says that some farmers in Central America place beehives around the perimeters of their property. But you just can't depend on Company Bee. "They aren't a reliable deterrent," McCrie says. "You don't know when they'll attack and when they won't."

Okay, so the birds and the bees aren't as flashy as biometric devices like retinal scanners that identify the capillary pattern at the back of your eyeball. "There's one possibility in a billion that two retinas are alike," says Jim Power, president of Martec Controls Systems. "And there are video scanners that match your face with pictures stored in their memory files." Power's company has created software to use with video scanners at building entrances. The system makes doormen obsolete; a watchman carrying a Watchman-like portable mini-TV screen can view the door from wherever he is on his rounds and buzz a person in by telephone.

Okay, so the Garden Club isn't going to replace Delta Force, even though plants like hemlock and narcissus are deadly poisons, and a kind of cactus called the jumping cholla actually drives its spines into your flesh. (I'm sure there's already a movement to foil natural security; it's rumored that the CIA's Covert Counter-Cackle-and-Crackle Command (CCCCC) has already found highly efficient methods of taking a gander or restoring silence when corn stalks at midnight.) And don't imagine a *Little Shop of Horrors* scenario in which vicious Venus'-flytraps take a bite out of crime. "Insectivorous plants don't have any characteristics that would make them useful for security purposes," says Matthew Hochberg, seventeen, a senior at the Bronx High School of Science and the person to whom the New York Botanical Gardens directs callers with questions about such greenery. He's been raising the plants since he was eleven. "Maybe a field full of Venus's-flytraps, which fold shut when one of six trigger hairs on the leaf is touched twice, could keep insects away," Hochberg says, "but that's all they can do. One of the largest carnivorous plants, *Nepenthes rajah*, which grows on Mount Kinabalu in Borneo, has been known to catch mice, frogs, and small birds in its pitcher-shaped leaves. Basically, mankind has nothing to fear from carnivorous plants." (Though nefariosos should, I think, tremble in dread of the day when there's sure to be a cyborg/plant hybrid called, maybe, Robocrop.)

Okay, so there isn't going to be a crack cadre of Green Bean Berets ("When those vegetables get steamed, watch out!"). But organic security devices have a simplicity and economy that many admire. Earl Gay is president of the New York firm Brooks-Gay Associates, which specializes in risk analysis and threat assessment. ("That's a pretty good threat, but not totally convincing. This threat stinks. Now, here's a threat with authority; I think this one—eeaaaarrrgghhhhhhh!") He asks that we meditate on the elegance of one venerable security device, the moat, with or without auxiliary crocodile capability.

The biggest problem with bushes as field operatives—their lack of communication skills—may someday be resolved. Researchers have found that drought-stricken plants emit ultrasonic noises—actually the fracturing of water pathways from root to leaf—that, if monitored, could tell farmers in arid countries precisely when they should water. Surely the next step is teaching P.T. to shout, "Freeze, scuzzball!" before he pierces the perp where it hurts. Some may argue that making a benign plant do our dirty work is a perversion of nature. But hey, it's a jungle out there, and sometimes you've got to call the copse.

Lifestyles of the Rich and Creamy

□ ■ □

Montezuma's real revenge may have been the cups of cocoa he offered Cortez and the Spanish soldiers who came to call in 1520. In those days, when a Mexican god-king served hot chocolate, he served *hot* chocolate—laced with ground chili peppers, not sugar. And there's no Aztec pictograph for marshmallows.

The Spanish hated the stuff, thick as honey, foaming, and dyed red so that it looked like blood. Joseph Acosta, a historian of the time, wrote of "a drincke which they call Chocholate, whereof they make great account, foolishly and without reason; for it is loathesome to such as are not acquainted with it, having a skumme or frothe that is very unpleasant to taste." The conquistadors shunned the beverage in favor of the golden goblets in which it was served. (Montezuma's little chocolate stunt may ultimately have backfired; Cortez double-crossed the captive chieftain, killing him even after he'd been ransomed, legend says, for a roomful of gold. "And that's for the skummey drincke!" the high-strung conquistador probably screamed, delivering the coup de grace.)

The conquerors were canny enough to tote home the beans of the tropical tree that two centuries later Swedish taxonomist Carolus Linnaeus would name *Theobroma* ("food of the gods") *cacao.* (Columbus was actually the first to bring back cacao, in 1502, after his fourth voyage to the New World, but King Ferdinand wasn't impressed.) Before long, someone thought to lose the chili and substitute milk, sugar, and vanilla. The recipe was a closely guarded secret until itinerant monks spilled the beans throughout Europe. In seventeenth- and eighteenth-century England, pub-like chocolate houses were the places to see and be seen; in fact, disgruntled beer and ale makers reportedly demanded legal restrictions on the competing brew.

By the time the English firm of Fry and Sons created the first chocolate bar in 1847, western civ was hooked. Though there's no pharmacological evidence that chocolate is addictive, people in thrall to the substance do odd things. Every spring, for example, over 500 scuba divers plumb California's Monterey Bay to hunt 350 chocolate abalones sunk in watertight containers. Several hundred thousand Americans annually attend chocolate weekends at resorts throughout the nation offering "come as your favorite chocolate" dances and fingerpainting with sweet umber goo. One such event at Miami's Fountainbleu Hilton featured the chance to dunk Don Johnson into a 600-gallon tank of chocolate syrup. Don wouldn't besmirch the Armani for any old foodstuff; you won't catch him plummeting into a pile of pemmican or a hunk of haggis. Cocoa-heads have been known to fly to Paris just for the hot chocolate at Angelina, where a real chocolate bar is melted in each cup (226 Rue de Rivoli, but keep it to yourself). They know facts like when the little white plume was added to Hershey's foil-wrapped kisses (1921). They—we—annually consume over two billion pounds—over ten pounds per person—of the food of the gods.

Little wonder that there was much hoo and ha over an article that appeared in a May 1988 issue of the *New England Journal of Medicine.* Two researchers at the University of Texas South-western Medical Center in Dallas became the Mr. Goodbars of science by demonstrating that stearic acid, a saturated fat plentiful in chocolate and beef, seems to lower cholesterol levels. (You'll recall that it was thought that all saturated fats—those found in red meat, butter, cheese, cocoa butter, coconut oil, and palm oil—*raise* cholesterol levels.)

Cardiologist Andrea Bonanome and internist Scott M. Grundy fed eleven male volunteers three different liquid diets, rotating regimens every three weeks. One diet contained stearic acid; one palmitic acid, the saturated fat in palm oil; and the other oleic acid, a polyunsaturated fat in safflower oil.

Compared with their blood cholesterol levels when the subjects were on the palmitic acid formula, total blood cholesterol dropped by 14 percent with the stearic acid formula and by 10 percent on the oleic acid diet. Levels of low-density lipoproteins (LDL, or "bad cholesterol"), the hematological hit men that carry artery-clogging goo, dropped by 21 percent in subjects on the high stearic acid diet and by 15 on the oleic diet.

For a minute there it was like the scene from Woody Allen's *Sleeper,* in which the hero, cryogenically preserved for 200 years, awakens in a future world where scientists have learned that hot fudge and deep fat are health foods. The Texas study made chocolate lovers rejoice; apparently it wasn't just okay to eat chocolate, but actually a lifesaver, to mix candy metaphors. And spokesbeefeaters Cybill Shepherd and James Garner would probably have danced a jig, if Garner hadn't been recovering from heart bypass surgery.

But before you could break out the champagne and Snickers, before you could break out—(naah, not really; chocolate beat the zit rap years ago)—the caveats flew. In an editorial in the *New England Journal,* doctors from the U.S. Department of Agriculture Human Nutrition Research Center noted that the liquid diet used in the study was otherwise low in cholesterol and that stearic acid might not perform the same way in real life, a nutty place where red meat and chocolate are usually part of a diet already rich in cholesterol. They also pointed out that beef fat contains palmitic as well as stearic acid, and that no one knows just how that combination works. The spoilsports concluded that current dietary guidelines, which call for Americans to reduce total fat and calorie intake, shouldn't change.

"The bad fats are butter and palm oil," says Bonanome, "but it would be dangerous to say chocolate lowers cholesterol—though it would make many people happy. The total amount of fat is higher in chocolate than in beef. As long as people eat lean beef, they'll probably be able to fit into the recommended guidelines for fat and calorie consumption. But I can't tell people to eat lean chocolate.

"And people tend to eat chocolate in addition to everything else. If I could tell them to eat chocolate *instead* of other things, it would be better."

Aww, fudge. That's the second time this decade that Science has toyed with the emotions of cocoa buffs. In 1982, two New York psychoanalysts, Donald F. Klein and Michael R. Liebowitz, earned brownie points by theorizing that the rush associated with passionate love is caused by a surge of an amphetaminelike brain chemical called phenylethylamine (PEA), which is also found in chocolate. When we fall out of love, they found, PEA production shuts down. People crave chocolate, the researchers suggested, because it gives them the same high as love or speed.

The pair made the connection while treating a group of love-addicted women, thrill-seeking types who suffered post-passion crashes during which they consumed mounds of PEA-laden chocolate. "Certain drugs called monoamine oxidase inhibitors, which regulate PEA, are especially good for people who don't respond to conventional antidepressants, including these women," explains Klein, now director of the anxiety clinic at the New York State Psychiatric Institute. "We were looking into the mechanism of how those drugs help such people. The chocolate part was an inference, really, a sideline that got more attention than any other work I've ever done. It was purely speculative."

"It was a silly theory," says Judith Wurtman, a research scientist in the Department of Brain and Cognitive Sciences at MIT. "There's not enough of the substance in chocolate to make any difference at all." We probably crave chocolate, she says, because it tastes good, has pleasant associations, and, like most carbohydrates, produces a sense of calm and well-being. Wurtman's work shows that carbohydrate consumption causes the pancreas to produce insulin, which sets in motion a chemical chain reaction leading to increased levels of the feel-good neurotransmitter serotonin. "That's why chocolate's satisfying—but no more satisfying than a potato would be," she says. "And a potato would do it faster; it's digested sooner because there's no fat. If you smelled chocolate and ate a potato, you'd probably get the same deep satisfaction." Raise your hand if you want a heart-shaped box of Tater Tots next Valentine's Day.

Chocolate may pack a slight high; it contains theobromine, a milder relative of caffeine that doesn't cross the blood-brain barrier. "Chocolate provides the same effect as coffee and a pastry," Wurtman says. "It perks you up and calms you down at the same time. Few foods combine such nice qualities."

Though some folks may feel physically incapable of resisting the siren song of a Sacher torte, chocolate is not addictive, Wurtman says. "It doesn't generate biochemical changes that cause adverse withdrawal symptoms when you stop eating it. You may crave it, but you may crave water, too, and water certainly isn't addictive."

She'd have a hard time convincing hard-core (or chewy center) chocoholics, the kind who don't stop feeding their heads long enough to thank the microbiologists, biochemists, and engineers who make their habit possible. Among the scientists behind bars is Larry Campbell, product development manager for Hershey Foods Corporation, who has a master's in dairy chemistry. One of his missions is to keep the world safe from fat bloom.

"Fat bloom is a physical appearance defect in chocolate," Campbell explains. "If fat in a candy bar melts, it migrates to the surface and leaves a gray film on top. Though there's nothing wrong with the chocolate, the appearance turns people off."

To prevent this scourge, chocolate must be tempered. "All fats, when they crystallize, take on polymorphic forms—each with a distinct shape and melting point," Campbell says. "Cocoa butter—the fat released by the cocoa bean—has five polymorphic forms. It melts quickly, almost at body temperature; that's why chocolate is so creamy. The way to prevent fat bloom is to cool chocolate below the crystallization point and warm it slowly through all five crystalline forms. This gives the smoothest and shiniest product."

Viscosity's another pain for the chocolate scientist. "If the chocolate's too thick, there'll be bubbles and holes and the consumer will think there's something wrong. If it's too thin, and you're enrobing—pouring it over a center, like nuts—it'll run off." Most consumers are too busy *dis*robing chocolate to ponder such problems. How many of us simply ignored the great syrup crisis of the seventies? "I spent a lot of years reformulating, improving, and developing new syrups," Campbell says. "We had a problem: the syrup would get very thick in the can on supermarket shelves. Or if it got hot, the syrup would separate. I was part of the team seeking the cause and correction." (Yes, they worked syruptitiously. Now be quiet.)

"We found that the starch in the cocoa was retrograding—the molecules were unwinding and re-reacting with each other to form a tangled matrix. We corrected the problem by adding an emulsifier that's used in bread." Campbell also works on developing different tasting products for different countries. Americans favor a cheesier-flavored chocolate than do Europeans, for example. Eastern Canadians choose a sweeter chocolate than do Western Canadians, and the Japanese prefer their chocolate syrup thick.

The Aztecs liked to grind corn into their chocolate, according to research anthropologist Jan Gasco of the University of California at Santa Barbara, who studies the cacao economy of Aztec and colonial Mexico. Those madcap Mesoamericans also used the flowers of the cacao tree to cure apathy and timidity, its bark for stomach trouble, its butter for burns, and its beans for money. Proving what every chocolate lover knows, unsaturated fats be damned: you can never be too rich or too rich.

Someone's in the Kitchen With Science

□ ■ □

I really didn't mind the cyanide-laced grape scare or the chickens tainted with heptachlor, a carcinogenic cousin of DDT. Bobbing for alar? What fun! A tuna-mercury casserole's okay by me as long as it's topped with crumbled potato chips, and I figure catsup detoxifies a hormoneburger. It was no big deal to learn that the Maillard reaction—the heated exchange between sugars and amino acids that browns cookies, breads, and meats—also produces substances that cause genetic mutations. I wasn't even alarmed when the *New York Times* suggested washing all fruits and vegetables in Ivory Soap. So what? I'll send my salads out to be dry-cleaned.

But then researchers at the University of Wisconsin announced that Cheez-Whiz is a potential cancer fighter. (It has to do with a group of fatty acids, in which the Whiz is rich, that inhibit some types of cancer in mice. The Wisconsin researchers also found high concentrations of the do-good fatty acids in grilled beef. You don't suppose that if the investigation had been done at the University of New Mexico we'd be finding out that blue corn tortillas are high in—naah.)

"The discovery of a new dish does more for the happiness of mankind than the discovery of a star," said Jean-Anthelme Brillat-Savarin, archetypal foodie and author of the classic *The*

Physiology of Taste. Oh yeah? And what does the discovery that the only thing left that's safe to eat is a toejamlike cheezfood do to the happiness of mankind? I was none too pleezed. Deciding it was time to toss my cookies in the ring and take a stand, I set out to find and spread (but not on crackers) some good news about science and food.

The Campbell's Soup Company recently commissioned a study called "2001: A Food Odyssey," and I suppose we should just be grateful that they didn't call it Foodbusters or Foodgate. It contains the mm-mmmm-good news that American and Japanese automakers will soon be equipping glove compartments with microwave ovens. By the turn of the century, a quarter of all drivers will engage in fast-track snacking or start cooking dinner on the way home. Designers are already creating packaging that can be opened with one hand.

I foresee a few problems. Imagine, for example, the traffic reports: "Watch for heavy congestion on the expressway, where an overturned turnover is blocking the left lane. And steer clear of a four-car heatup on the Interstate. . . ." There'll probably be stiff penalties for defrosting cheesecake in a blintz zone, and pit stops will take longer: "Yeah, your oil's down a quart, the right rear tire's low, and you could use a coupla Eggos and some fish sticks."

But it might be fun. Maybe food items will be customized, like Chevre Rolet, Alfalfa Romeo, Subarbecuru, Rack of Lamborghini, or the zippy orange drink astronauts love, MusTang. And I can't wait for the Pillsbury 500.

Meanwhile, the folks at Boston-based Steve's Ice Cream recently decided they wanted to create a microwaveable hot fudge sundae, one packaged in a way that allowed the fudge to melt but kept the ice cream cold. Was Science up to the challenge? It has never, after all, explained how socks travel from washing machines to another dimension, or why your hair, after looking freeze-dried for weeks, does just what you want it to on the day you're having it cut. But Steve's research and development people had faith that Science could lick ice cream any day.

"We bought the idea from a chocolate company," explains Emily Skoler, Steve's special products coordinator. "They'd come up with the idea of enrobing the ice cream in fudge, so that the fudge protected it from the microwaves. But if you didn't take the sundae out exactly on time—if you were off by a second or two—it didn't work."

The researchers decided to fiddle with the basic ice cream formula of water, sugar, and butterfat. They substituted corn syrup for sugar, since it heats more slowly, and added more water for extra iciness. That helped, but the product was gritty. A grit-fighting additive made the stuff taste like hell frozen over.

So they worked on packaging. Protective foil petals inside a plastic cup seemed promising but allowed microwaves to sneak in and arouse the frigid ice cream molecules. Finally, after nearly a year of work, the team hit on the perfect plan: They placed the ice cream in a completely foil-lined cup and suspended the fudge above it in a plastic dome. When zapped, the fudge melts and drizzles over the still-chilly ice cream.

Okay, it's not a cure for cancer, but it's still a technological triumph worth savoring. And it may have important applications. Suppose, for example, a person can't decide if she wants to be cremated or cryonically preserved after death. Bingo! Thanks to Steve's, she's got the best of both next worlds. The sundaes, which sell for $1.50, are being test-marketed.

Why, you might ask, couldn't the fudge be sold—and heated—separately, eliminating the need for all this toil? "Why do those soups come in Styrofoam containers you can heat up and eat from?" asks Skoler. "In this case, it seems to matter that people don't have to dirty a scoop or rip open a separate package. The consumer seems to think that's important. And the idea of ice cream and fudge being microwaved together is exciting in itself."

Calvin Schwabe, on the other hand, hopes that someday we'll all scream for mice cream. He'd also like to put the ant back in antipasto—see recipe below.

Schwabe, an epidemiologist at the University of California at Davis School of Veterinary Medicine, thinks Americans can't afford to turn up their noses at alternative sources of protein. "Our prejudices are contributing to the world food shortage," he says. That's why he's written *Unmentionable Cuisine,* a collection of rather arresting recipes gathered during thirty years of working in third world countries for agencies like UNICEF and the World Health Organization.

Who wouldn't succumb to dishes like Broiled Puppy with Sweet Potatoes, Earthworm Broth, Fish Sperm Crepes (guess who's coming for dinner!), Brain Fritters, and Pig Uterus Sausage? Thumbing through Schwabe's cookbook, you almost think you could expand your definition of good grub to broiled grub. (Of beetle. In soy sauce.) Almost.

"I don't include any ecologically threatened species," Schwabe points out. "I've stressed using eating as pest control. I found forty French recipes for starlings, for example. And in parts of Thailand, rice rats are a big part of the diet." Among his pest bets: Grilled Rat Bordeaux Style (no Ratatouille, though), Starlings En Croute, and Mice in Cream.

Schwabe's sure the reluctant can be won over. "At my annual Unmentionable Dinners—a tradition that ended three years ago—guests were dubious at first, then enthusiastic. I've served starling, crickets, sea urchin gonads—poor man's caviar. And turkey testicles; people liked those." Everyone had a ball. And why not? Could turkey testicles be any worse than, say, Ritz Mock Apple Pie?

Schwabe tests many of his recipes at home; his motto seems to be, Why order out when you can eat innards? His wife and two children serve as guinea pigs—and have been served guinea pigs, a South American delicacy.

Oh, here's that recipe for Chindi Chutney, made from red ants: Simply collect ants in leaf cups and place directly into hot ashes for a few minutes. Remove ants, grind into paste, add salt and chilies, and bake. "It is said to have a 'sharp, clean taste' and is often eaten with alcoholic drinks," Schwabe writes. But who has time to home bake ants or hand mix drinks anymore? What this country needs is a microwaveable package that keeps the red ants hot and moist, and the martinis cool and dry. (I'm told, incidentally, that at the Explorer's Club in New York City, initiates are served martinis garnished with a goat's eyeball instead of an olive. Here's looking at you, kid!)

We know that drugs can turn people into vegetables. Now a researcher has discovered that vegetables can turn people away from drugs, specifically tobacco snuff. The American Cancer Society reports that 30 percent of all adolescent males are up to snuff, and that the number of female snuff-dippers is rising. A pinch between cheek and gum delivers twice as much nicotine as a cigarette, speeds heart rate, and elevates blood pressure; over time, big dippers risk coronary disease, stroke, and cancerous lesions. And snuff's as addictive as morphine and cocaine. "It's a lot easier to stop smoking than it is to stop dipping," says Dr. Richard Glass, chairman of the oral pathology department of the University of Oklahoma Health Sciences Center. He's testing an unusual snuff-cessation program aimed at students. What's his secret weapon?

Well, broccoli, of course. Glass has found that ground raw broccoli heads make an excellent, safe substitute for snuff.

Scientists have known for some time that broccoli contains cancer-fighting chemicals called indoles, as do other members of the cabbage family. Now it seems that something in broccoli kills the hankering for snuff. Glass isn't sure what, though. "It's partly the consistency, and perhaps because broccoli has enough nicotinic acid to appease the craving," he says. "There also seems to be a poultice effect; people seem to have a lessening of lesions when they use broccoli or an herbal snuff substitute. Perhaps it's the placebo effect. I convince people if they use broccoli they can kick snuff, and they do."

As if the broccoli cure weren't enough to enhance the cachet of produce, it now turns out cucumbers are a lot cooler than we thought. Their leaves contain an enzyme able to transform cholesterol into coprostanol, a harmless fatty acid that passes through the body unabsorbed, says Iowa State University biochemist Donald Beitz.

You can't just pop a cucumber with your omelet, Beitz cautions. His idea is to treat food with the enzyme cholesterol reductase before we eat it, so that cholesterol has turned to coprostanol before we stuff our faces. "Visualize a salt, pepper, and cholesterol reductase shaker on your table," Beitz says. "You'd sprinkle some of the enzyme on burgers or sausage or scrambled eggs before you fry them. We're trying to keep the incubation time at a convenient minimum, about five or ten minutes. Milk could be treated the same way." Beitz is looking at alfalfa, peas, corn, soybeans, oats, and mushrooms as sources of reductase. Don't look for the shakers in the miracle case of your supermarket. Even if his scheme works, Beitz says, it'll be three to five years before things go better with cuke.

No smorgasbord of good news about science and cuisine would be complete without mention of the world's only independent food museum. Visiting the Potato Museum in Washington, D.C., was like getting hit over the head with a four-foot-high Belgian potato masher—just that eye-opening. The museum had such a monster masher on display, as well as 2,000 other artifacts relating to the scientific and social history of the potato, fourth most popular staple (after wheat, corn, and rice) and consuming passion of Tom and Meredith Hughes, founders and curators.

Housed from 1983 to 1990 in the basement of the Hughes's rented Capitol Hill Victorian, the collection began fourteen years ago as a classroom project at the International School of Brussels, where Tom taught. (The Belgians take their potatoes seriously, you learn at the museum; they invented so-called French fries and make, arguably, the best in the world. The secret: fry twice in ox fat.) The Hugheses became ardent potatophiles, deciding to devote their free time to championing *Solanum tuberosum,* let the chips fall where they may.

"So little attention is paid to the role food has played in human development," says Meredith, a free-lance writer. "The directory of the American Association of Museums doesn't even list food among its hundreds of categories of subjects." Adds Tom, who teaches fifth grade at the Potomac School in McLean, Virginia, "Where are the people preserving the history of the world's foods? It's up to us, as archivists and advocates of the potato, the world's most important vegetable." A field of spuds, it seems, produces more energy per hectare per day than a field of any other crop. I'd never heard anyone couch potatoes in quite these terms.

The museum was underfunded—it received a small grant from the International Potato Center in Lima, Peru, and the Hugheses anted up the rest—but not half-baked. In fact, it was everything a museum should be—imaginative, intimate, seductive. You could start with the Inca (potato-motif pots; ancient foot plows like the kind still used in the potato's Andean birthplace) or with the Irish (press clippings from the 1845 famine, a preserved potato suffering from the blight that changed Ireland's demographics). A slide show offered a look at some of the world's 10,000 varieties of potatoes, including a gnarled Andean specimen so unappealing and unpeelable that its name in Quechua, an Indian dialect, translates as "the potato that makes the daughter-in-law weep." There was a display of nineteenth-century autochromes, the first practical color photographs, made in the early part of the century with dyed potato starch granules, and of oddities like potato ice cream, cigarettes, hand lotion, and paper. Naturally, Dee Dee Sharp's 1960 hit album "It's Mashed Potato Time" was prominently displayed. My favorite exhibit was an evolutionary rogues' gallery of the Potato Head Family, including a rudimentary pre–World War II protopotato, the fondly remembered archetypal fifties version and the current incarnation, a bulbous plastic boobtuber who strikes me as hideously maladaptive.

The museum's archives contained copies of nearly every book ever published about the potato and served as an informal clearinghouse for potato research. When I visited, for example, the pair had just been talking with spudniks at the University of Wisconsin (the Potato Department is right next to the Center for the Advanced Study of Anticarcinogenic Snacks) who are experimenting with growing potatos in zero gravity for NASA. The diminutive museum was, in other words, no small potatoes.

Alas. Last summer, the Potato Museum was forced underground. The Hughes' landlady took her house back and the couple has moved to Great Falls, Virginia, with their five-year-old son Gulliver—yes, I imagine he's been called a tater tot—and their dog Tato, too. The museum's holdings are stored in the basement, awaiting rescue by corporate or community sponsorship. Several American towns and Shannon, Ireland, have expressed an interest in the museum, but so far no one's come through. This setback hasn't taken the starch out of the Hugheses, however.

"Everything's still potato-oriented with us," says Meredith, who continues to edit the museum's quarterly newsletter, *Peel-*

ings. Tom is currently one of five guest curators preparing a Smithsonian exhibit called "Seeds of Change," commemorating Columbus's voyages; he'll chronicle the potato's journey from South America to Europe with the conquistadors in the late 1500s and back to North America in 1620. And the pair is campaigning to nickname Washington, D.C., the Hot Potato. "Somebody's handling one every day there," Tom says.

Finally, remember when the international scientific community was all het up over the announcement that University of Utah chemists claimed to have created fusion at room temperature, using a jar of heavy water and an electrical circuit to create energy? The excitement proved premature, but such an achievement would have been nice, because it would have meant an endless supply of cheap energy without the fuss and muss of heating your jar until it's as hot as the sun, which is powered by nuclear fusion. But what none of the news reports told us is what kind of jar the scientists used. Suppose we wanted to try this at home?

According to Mark Andersen, a postdoctoral associate working on the project, the jar isn't a jar but a Dewar's tube, which, he says is actually one tube within another, something like a thermos. "Ours was an inch and a half to two inches in diameter, depending on the size of the palladium rods we used, and handblown by the chemistry department glassblower." There you have it.

You know, even if this cold fusion thing is someday hailed as a breakthrough ranking with the discovery of fire or the Slinky, some of us will still want energy like Grandma used to make. Sure, reaching fusion levels requires lots of preparation time, plenty of counter space, and hundreds-of-millions-degree temperatures. It can get tricky if you're trying to prepare a hot fudge sundae at the same time. But if you can't stand the heat, get out of the kitchen.

MINERAL

□ ■ □ ■ □ ■ □ ■ □ ■ □ ■ □ ■ □

Technology sometimes makes people nervous, probably starting with the first Paleolithic purist who said, "Stone tools? No way! Too dehumanizing!" Plato, in his *Phaedrus,* tells us that when Thoth, the Egyptian god of wisdom, invented the alphabet, the legendary god-king Ammon wasn't impressed. "This discovery of yours," Plato quotes him as saying, "will create forgetfulness in the learners' souls . . . they will trust to the external written character and not remember of themselves . . . they will appear omniscient and will generally know nothing."

So far, the alphabet hasn't been yanked off the market for causing brain damage. So don't worry about Hide-a-Gun and Mice-Cubes.

□ ■ □

Widget Goes to Washington

□ ■ □

Okay. Perhaps U.S. Patent Number 4,632,062 will never have quite the impact of, say, U.S. Patent Number 223,898. Still, the Squirrel Twirl, brainchild of retired Indiana barber Jack Hubbard and one of fifty new inventions on display at the sixteenth Annual Inventors Expo held in Arlington, Virginia, is spiritual sib to Edison's electric lamp.

Both inventors had a mission, you might call it. Hubbard's was to keep the squirrels the heck away from his bird feeder; Edison's, to illuminate the world. Each solved his problem elegantly. Edison created the carbon filament. Hubbard drove a stick into the ground and screwed another one to it, leaving enough play so that it could spin propeller-fashion. At the top of the spinnable stick he nailed an ear of corn. When a squirrel jumped for the corn, the animal's weight swung the stick down. At first, the startled creatures jumped off. "But then they liked the ride," Hubbard reports. "I didn't mean for it to be entertaining, but now I incorporate that into my description: 'useful and entertaining, with 360 degree acrobatic action.' Any nut could have invented it," he says, probably not for the first time, "and one did!"

"The more original the discovery, the more obvious it seems afterwards," Arthur Koestler said in *The Act of Creation.* "Why didn't I think of that?" visitors to the Expo—nearly

10,000 over a spring weekend—kept saying as they strolled about, so diverse in age, sex, race, color, and creed (actually, I'm only guessing about creed) that they could have posed for Norman Rockwell's *The Golden Rule.*

What I was saying to myself, as I quickly reconnoitered the exhibition hall, checking out Catch-A-Cap, Redi-Ratch, Turbo-Coil, Scan-A-Line, Wee-Changer, Hitch-Mate, Lug-Loose, Swift-Answer, and Hide-A-Gun, was that I wish I'd invented the hyphen. If someone had thought to patent that petite but puissant punctuation mark, apparently so popular among innovators, she'd be super-rich today.

A person could use several different methods for organizing a browse through the exhibits, selected by a U.S. Patent Office committee to represent a cross section of the latest among the 4.7 million or so American patents granted since 1790. George Washington signed the bill creating the system; the first registered patent went to Samuel Hopkins of Pittsford, Vermont, for an improvement in "the making of Pot ash and Pearl ash by a new Apparatus and Process."

A thematic approach seemed interesting. So first I explored items declared New, like John W. Fenton's New Mechanical Principle for an Engine (patent no. 4,589,388)—it has sort of yin/yang-shaped twin rotating combustion chambers, but no pistons, connecting rods, crankshaft, camshaft, or engine block—and David Sterner's Direct Opposite Reverse (patent no. 4,572,622), the First New Geometric Shape in Six Thousand Years. The young Pythagoras of Portland, Oregon, created this sphere-cube combo by halving a square-edged, domed Plexiglas aquarium hood and then twisting and reconnecting the halves. A Direct Opposite Reverse–shaped photographic lens, it seems, can enlarge images in the camera. There was also some talk about "the science of trinities that equate infinite balance," but I didn't understand a word of it.

Dr. Youssef's Toothstix (no. 4,576,190), on the other hand, weren't hard to fathom: they're A Totally New and Different Material for Brushing the Teeth. "It is a polyurethane foam which works in a very different way from the bristle brush," explained the Palm Beach pathologist with a big, clean smile situated under a panama hat and over a broad expanse of flowered tie. "It wipes the plaque off the teeth, especially when assisted by a detergent such as toothpaste." He modeled its size, shape, heft, and disposability, he says, after the stick used for corn on the cob at Kentucky Fried Chicken.

My next foray was planned around a Sports and Recreation theme: this included a "bow draw sighting device" (no. 4,179,613) for archers that beeps and blinks when your draw is perfect (I would have called it William Tells All); Autotroll (no. 4,603,499), a dealie that holds your fishing rod and waves it slowly back and forth; and the Strong Shifter (no. 4,548,092), a clever way to change bicycle gears using easily installed handgrips. And there was Catch-A-Cap (no. 285,141), just like an ordinary baseball cap, except for the four finger loops atop its padded crown. Should a high pop foul come toward you in the stands, see, you can whip Catch-A-Cap off your head and use it as a glove. "If you'll notice," says inventor Kenneth Whisman of Fairfield, Ohio, a longtime Reds fan, "when a ball hits down in the seats, the fans fight over it. I thought there might be an easier way. At first, I had a complete glove on top of this, with a thumb loop. But I would have had to make it for left-handers and right-handers. I was all the way to a patent attorney before I came up with just these straps." You may be busy formulating some cruel joke about how Red Sox first baseman Bill Buckner shoulda hadda Catch-a-Cap during the sixth game of the 1986 World Series, when he missed a crucial ground ball, thus gift wrapping the game for the Mets. Well, forget it. "We don't recommend the Catch-a-Cap for daisy cutters, because those are too hard. We're talking about the one that arches into your glove."

Another theme I pursued was Stuff Having To Do with Haircuts. Robocut (no. 4,602,542) was invented by Alfred Natrasevschi, an electrical engineer from Romania who now works for Hewlett-Packard in Fort Collins, Colorado. It looks something like a blow-dryer, attaches to your vacuum cleaner, and costs $34.95. Suction holds a chunk of hair straight so a rotary blade can whack it off. Natrasevschi, an attractive guy, says he uses it on his own hair, which does a Mickey Rourke kind of thing that's a bit discouraging. But none of the thirty or so men, women, and children who had a demo haircut punched him or cried.

At a nearby booth was displayed the equally intriguing but cheaper Cut & Style (no. 4,619,280), a $7 transparent plastic sheet printed with a grid and curving lines; it snaps around a person's head and the hair is combed over it so the stylist can cut by the numbers. "Are you the inventor?" I asked the woman in the booth. "I have to confess, it's me," she said. "I have high, high, education, and when I come to this country—"

"Where are you from originally?"

"My boss say no personal questions. Is American company and I am American for a long time." I observe her name tag. You can't kid a kidder, Ludmila Yampolskaya, especially when her grandfather's from Kiev. Yampolskaya, once managing editor of a Russian scientific journal, says Cut & Style is getting mixed reviews. "In countries where hairdressers' fees are fixed [a euphemism for behind the Curling Iron Curtain], hairdressers like it. Here they consider it a threat." The recent reforms in Eastern Europe may be good for Freedom & Democracy, but bad for Cut & Style.

My next theme: Things You Might Find in the Classified

Section of *Soldier of Fortune* Magazine. "Do you have any particular highways or bridges you'd like to take down?" chemical engineer Richard Snyder kindly inquired. If I do, he's got the baby for the job: the Natchez Remote Radio Blasting System (no. 4,576,093). For a mere $2,700, it stores 531,000 possible touch-tone secret detonating code combinations, provides unlimited standoff distance from the blast—something you probably look for in an explosive device—and minimizes premature firing, a feature Masters and Johnson are said to be studying.

Hide-A-Gun (no. 4,570,890) is a kind of metal mini-gunrack for the home. "It's a means of keeping a gun out of sight but ready for quick use," explained inventor E. E. Lohn. "Like, see the one under there?" He pointed to a .38 under a shelf, just beneath an "I ♥ Grandpa" mug. "I have a Hollywood headboard, and I used to lie in bed and look up there and decide that was the place to have my .25 automatic. The prototype was a block of wood with a two-inch piece of five-sixteenths copper tubing. In fact," he chuckled, "my wife still has a wooden one on her side of the bed!" He'd better hope she doesn't wake up some morning in a "I 🔫 Grandpa" kind of mood.

It was time to check out Inventions That Make the World a Better Place, among them a souped-up, comfy, easy-to-control wheelchair (no. 4,614,246); a timed, rotating pill dispenser (no. 4,572,403); and, from emergency physician Alfred Frankel of Florida, a safer, guided system for inserting endotracheal tubes to clear a patient's airways (no. 4,672,960). Al Chase of Freedom, Maine, worries about the eleven million babies of diaper-wearing age whose caregivers need to change them in public places. Hence Wee-Changer (no. 4,613,996), a 30-inch-high platform that pulls down from the wall and has a disposable covering, a Velcro strap to secure the kid, and various hooks and recesses for baby rubble.

Then I simply wandered. One booth displayed a large white felt banner appliqued with big pink roses. "Sudden a thought came like a full-blown rose" it said, a quote from Keats. The people at Rose Hill Farms aren't the inventors of the much-sung blossom itself—that, of course, was Pete Rose—but they have patented several miniature versions. Explained Pat Berlen rather explicitly, "We take pollen from a father and place it on stamens of a mother with characteristics we find desirable and would like to pass on to a seedling. We evaluate the seeds, and if they have excellent qualities—fragrance, growth characteristics, disease resistancy, color—we patent them." Plants have been patentable since 1930, but the world's first animal patent wasn't granted until 1988, to Harvard University, for building not the proverbial better mousetrap (see page 145), but a better mouse: a genetically altered rodent that will help them study cancer.

Trademark Office attorneys Cathy Dobbs and Frank Hellwig, there to demonstrate the agency's computerized system, obligingly checked to reassure me that the Grateful Dead logo—the garlanded skull—is duly registered. Hellwig said that there are several reasons a trademark can be refused registry: if there's already a similar one, for example, or if it includes a word someone else might want to use, like *computer*, or if it contains "scandalous material." He recalled a recently published opinion in the trademark report of the *U.S. Patent Quarterly*. "The proposed mark was Booby Trap for brassieres. It was considered offensive and couldn't be registered."

If you wanted to file for a patent—from the Latin *litterae patentes*, "open letters," after the missives given by medieval kings granting rights and privileges—you'd need to search through the twenty-eight million documents of the U.S. Patent Office to make sure no one had beaten you to the punch (no. 67,813, a paper cutting and punching machine patented 1867). And until the computerization of the archives—begun in 1984 and plagued with glitches and cost overrides—is complete, you either have to do it by hand or through a number of commercially available data bases covering more recent patents. Jason Crawford, eight, of Silver Spring, Maryland, was taking advantage of a free demonstration by the Durwent Company, a patent search service that usually charges $288 per connect hour. Jason wanted to find out if anyone had patented a portable folding stroller ramp like the one he's invented.

"I was at a fair with my mom," he said, "and there were a bunch of things in the stroller—not my baby brother—and I was trying to get it down some stairs and the whole thing tumbled over," he recalled, "so I thought of the ramp. It'll be light strips of metal that roll up.

"I've also invented something called spray soap that will give new meaning to the term liquid soap because it's so thin you can spray as if from a spray bottle. And I invented a mini-fan, with a microchip inside." Jason thinks Inventors Expo is "neat" and suggested I go look at the kids' inventions—there was a booth displaying the winners of the Weekly Reader National Invention Contest, and another from Project XL, a Patent Office program. "I thought a neat one was a watch that talks to you—it calls out the hours and minutes."

The Talking, Teaching Wristwatch was the $500 grand prize in the *Weekly Reader* contest's secondary school division. Inventor Ronetta Williams, a fifteen-year-old ninth grader from Pittsburgh's Schenley High School, says she was inspired by a little neighbor who couldn't tell time.

Maybe Williams will one day crack the National Inventors Hall of Fame, whose seventy-one laureates are celebrated in portrait and audiotape in the lobby of the Patent Office building, right above our heads. Though in 1988 computer wizard An Wang became the first Asian honored, the Hall of Fame is pretty feeble in the affirmative action department. "No Gurls Alowed," the first (posthumous) inductee, Thomas

"Spanky" Edison, might as well have written on the clubhouse door. Okay, I see why Enrico Fermi gets in for inventing the nuclear reactor, but Marie Curie doesn't for discovering radioactivity. Discovering and inventing aren't exactly the same; Einstein didn't make the team, either. But how come Eli Whitney is honored over Catherine Littlefield Greene, who should have told him to keep his cotton pickin' hands off her gin? Furthermore, why is Luther Burbank hailed for messing with the peach, but not George Washington Carver for improving the peanut? Which has contributed more to the American Way? For the answer, just sing this slightly altered anthem: "Take me out to the ball game/Take me out to the crowd/Buy me some *peaches* and Crackerjack. . . ." Uh-huh. I rest my combination back scratcher, potato peeler, and case.

On March 1, 1991, the first woman was inducted into the Inventor's Hall of Fame! Gertrude Bell Elion, 73, who won a 1989 Nobel Prize for Medicine, was recognized for creating drugs to fight leukemia, septic shock, and transplant rejection. "I'm happy to be the first woman in the Hall of Fame," she said. "But I doubt I'll be the last."

Putting Your Money Where Your Mouse Is

□ ■ □

So how do you want your mice iced? Scan a list of recent additions to the 4,000 mousetrap patents registered since the first simple snapper in 1838 and you'll find a provocative armamentarium. You can now send rodents to the Big Cheese in the Sky via gas chamber, guillotine, harpoon, electrocution, explosion, drowning, overdose of liquid nitrogen, asphyxiation, or heart attack. All that's missing are wee cement overshoes and a Tomahawk missile tipped with weapon-grade cheddar.

One hundred and thirty-four years after Ralph Waldo Emerson sort of said, "Build a better mousetrap and the world will beat a path to your door," people are still trying to perfect the archetypal invention, some form of which seems to have existed since the Babylonians. Every year, fifty to one hundred new mousetrap patent applications are filed, according to Nick Godici, the U.S. Patent and Trademark Office's supervising examiner for fishing, trapping, and vermin destroying. Of those, 65 to 70 percent eventually receive patents.

People build better mousetraps for the same reasons they climb mountains and eat Rice Krispies squares—they're there, crying out to be improved upon. Other motivators: Dreams of wealth, for one thing; Americans spend $300 million a year on professional rodent control. (Amateur rodents frolic unmolested.) The exhilaration of challenge. And perhaps deep personal grudges. After reading some of these patents, you'd think it was the rodent, not the serpent, that caused property values to drop in Eden. Parental discretion advised: Parts of the following contain scenes of explicit violence and words like *mash, flatten,* and *mutilate.*

If, like me, you can't even sing "she cut off their tails with a carving knife" without getting all misty, you'll be pleased to know that several of the very newest traps are humane catch-and-release affairs, perhaps heralds of an age of kinder, gentler mouse removal.

But not U.S. Patent Number 4,669,216, "Apparatus for Trapping and Disposing of Rodents," which wins the Torquemada Award for Extreme Thoroughness. Let the official patent documents speak for themselves: "When the presence of the rodent is sensed, a harpoon is fired downward through the rodent. A top plate then moves downward toward a bottom plate, to flatten the body of the rodent. Electric heating coils then incinerate the body of the rodent. After the body of the animal has been sufficiently incinerated, the trap is automatically reset." What? Doesn't it gnash the ashes? Accompanying this description is a drawing of what looks like a miniature version of the machinery that ingests Charlie Chaplin in *Modern Times.*

Call me Ishmael, but wouldn't the harpoon alone ("Thar she squeaks!") be sufficient to ahabilitate the critter? The paper trail to the man who could answer that question, inventor Ted M. Moss, dead-ended in Fort Worth, Texas. My guess is certain friends of the late Carlo "Chisel-like Incisors" Gambino offered to nibble Moss's lungs out, causing him to seek asylum within the Federal Inventor's Relocation Program.

A good brand name for Charles Shurden's trap (no. 4,766,692) might be Tail of Two Cities: "The novel mousetrap is of the guillotine type," the official description goes, "comprising a guillotine or bladelike member which strikes the neck of the mouse to inflict fatal injuries without severing the head from the body, thereby allowing easy cleaning of the trap and disposal of the dead mouse." The mouse—let's call him Sydney Carton—enters a sort of plastic box containing the guillotine mechanism. Sydney takes the bait from a lever that then pivots, releasing a striker that in turn releases the guillotine, which slides downward and does its job. Rabble and knitting needles extra.

The Marbles Counter-Weighted Repeating Mousetrap (no. 4,154,016), invented by George Millard of Cardiff, New Jersey, seems like a Sunday school picnic in comparison. It's a seesawing, troughed platform that clips onto any old bucket partially filled with water. When a mouse moves forward, lured by bait smeared on the front end of the platform, the thing teeters under the weight of the animal. Marbles at the rear of the trough roll forward, causing that sucker to tilt so fast the mouse can't escape; it plops into the bucket and drowns. The trap moves back into position for the next victim. And I guess when you have enough, you make miced tea.

Ten years ago, inventor Daniel Reyes created his Mouse Trap With Bait-Holding Tilt Tube (no. 4,154,016), a device as yet unsurpassed in Rube Goldbergian splendor—and as yet unmarketed, according to all available information. This is the only mousetrap that may someday be optioned as a miniseries.

Imagine, if you will, a small, plastic version of the mothership in *Close Encounters of the Third Kind,* about two feet in diameter. Further imagine that this multi-portholed, split-level edifice is, as the documents put it, "illuminated inside when entered by animals and which has windows so a person can see the animals inside." Imagine you are the person watching. What you see is a sort of mouse Carnegie Deli: Cheese and bacon are suspended from three hooks. Mice, lured up a gangplank and through the door by still more bait outside, race for the food but step onto trapdoors that plunge them into a lower chamber filled with water. The pivoting of the trapdoor closes an electric circuit, turning on interior lights so you can witness this lethal water ballet. The mice in the very charming illustration seem to be enjoying themselves.

My nominee for the mousetrap of the nineties, which promises to be the Age of Ambivalence: Zap-a-Mouse with CBL-2 (no. 4,805,340), offering consumers the option of dispensing life or death as the mood strikes them. On the market less than a year, it's a simple adhesive trap—the creature's immobilized in goo—but the bait pellet contains what is essentially mouse valium: CBL-2, carbromal, is a nonnarcotic animal tranquilizer. However you choose to deal with your prisoner—whack it upside the head, let it starve, or peel it off and free it—you're dealing with one very mellow mouse. If Mickey and Minnie ever check into the Betty Ford Clinic, you'll know why.

"Zap-a-Mouse is a better mousetrap, an improvement on plain glue traps," says Ronnie Becker, vice president of Tamby Chemical Company of Brooklyn, New York. Its inventor, Becker's late father, Jerome, was inspired, she says, by learning that animal rights activists were protesting snap and plain glue traps. "When mice get caught in a regular glue trap, they scream all night or mutilate themselves, gnawing their feet to get free." The Becker solution? "A light glue and bait with a tranquilizer that calms the mouse and keeps it from struggling or mutilating itself. Then the person can either destroy the mouse or set it free. Our research shows that ninety-nine times out of a hundred, the mouse eats the pellet and is calmed down." The Humane Society accepts the trap, Becker says, though it won't grant its seal of approval unless the package explicitly states that the mouse can be freed and released. "I can't do that," Becker explains. "If one person got bitten, I'd be sued out of business.

" 'Saturday Night Live' did a spoof on us," she reports. "Only they changed it from mice to rats. We don't put tranquilizer on our rat traps." Snare the rodent, spoil the child: "You'd have to put enough on to knock out a kid."

Ken Bernard's Mice Cube (no. 4,787,170), manufactured by Pied Piper International of Plaistow, New Hampshire, also lets the trapper make life-or-death decisions. Mice enter an opaque plastic box baited with peanut butter behind a door that swings one way. The consumer can turn the trap upside down so that the door flops open and the mouse is released on his own recognizance, or said consumer can leave the trap as is until the mouse asphyxiates.

John Fodor of Morgantown, West Virginia, received a patent in April of 1989 (no. 4,819,368) for a trap that offers a karmic twofer: "It's humane and it recycles aluminum cans." Fodor's design calls for enlarging the can opening and adding a small door connected to a baited rod that extends into the container and is held in place by a taut rubber band. When a mouse takes the bait, the rubber band is released and the door slams shut. Fodor won't discuss the price or name yet, but he plans to start selling his traps, made from scavenged cans covered with wood-grain Contac paper, very soon. "I hope to market it not only as a functional mousetrap, but as a novelty," he says. "I'm hoping it ends up on everyone's desk as a conversation piece. This is the ultimate. The better mousetrap has been built."

Melvin Melton of San Clemente, California, is also not shy about putting his money where his mouse is. "I don't think there's any question about it—this is the better mousetrap they talk about," says Melton of his Trap-Ease (no. 4,578,892), a square tube about 6 inches long, bent at one end. The mouse runs through an open one-way door, tilts the tube, and shuts the door. "Most other traps mash the mouse, and ectoparasites—fleas, ticks, whatever—jump off. More people have been killed by fleas than by all the wars in the world. That's how rats carried bubonic plague."

Melton, a retired dude ranch owner, patented his trap in fifty foreign countries, too, a wise move: A Malaysian factory turning out a larger version of the Trap-Ease—"terrific for rats and mongoose!"—just got an order for 400,000 of them from Rentokil, a huge British pest control company.

My personal favorite among recently patented devices is not actually a mousetrap, but the Mouse Trap Game (no. 4,508,352), created by William Johnson of Glendora, California. It's a game of chance employing a random selector—like the ball on a roulette wheel—that happens to be a live mouse. The animal is let loose on a 5-foot-square game board; if he heads down the "selector aperture"—hole—that you've bet on, you win.

Johnson says he'd remembered seeing something similar at a carnival when he was a kid; he made his for a festival thrown by his parish church. Now he rents the games for $500 a weekend, to charity fundraisers only; they're a huge success. He catches his own mice in a humane trap called Tin-Cat and makes sure they get plenty of vacation time. "The only problem I ever had was when we drew such a huge crowd the professional carnival people were jealous and called the sheriff, saying we were gambling," Johnson says. "The district attorney laughed and wouldn't file charges."

(I hope this spawns a whole genre of games with live animal playing pieces; raised stakes could make familiar games far more interesting: "Miss Scarlet in the billiard room with the pit viper!" The trend may already have started. On a recent trip to the Berkshires, I bought a chance to win $3,000 in a Meadow-Muffin Raffle: A cow was let loose in a field marked off in numbered squares. If it deposited a pattie in the square the number of which corresponded to the one on your ticket, you won the dough. But I digress.)

Has the better mousetrap been built? It depends on what you mean by better, and on whom you ask.

Ask the folks at the Woodstream Corporation of Lititz, Pennsylvania, the largest manufacturers of the classic Victor snap-trap, the kind frequently found dangling from a cartoon character's nose. In various corporate incarnations, Woodstream has held the patent for the basic trap since the turn of the century. "I don't see how you can improve upon something that's so simple and effective," says Kitty Miller, manager of marketing services. "Though we said we'd built a better mousetrap when we came out with the Easyset." That's a version of the Victor preset with Swiss cheese–flavored plastic bait. "Some people object to the mess," she adds, meaning bloody little corpses. "But more people object to mice lurking in their houses, spoiling food, carrying disease, and leaving their little leavings."

Then ask again, because—just as the same inner-city block can produce both a hit man and a priest—Woodstream is also the biggest American manufacturer of humane traps. Their research and development department came up with the groundbreaking Havahart humane trap, one of the first mass-marketed plastic boxes with one-way doors. Still, the snapping trap outsells it. And the higher the consciousness, the higher the price: the Havahart retails for about fifteen times more than a Victor.

Ask supervising inspector Godici, and you get another answer. "The qualities of a better mousetrap are the qualities of any invention worthy of a patent," he says. "It must be new, useful, and unobvious. It's a legal determination. The best mousetrap is the one that becomes commercially successful and allows the inventor to enjoy a profit.

"One that's been very successful is the simple tray of glue used to catch mice. That was invented by Stanley and Benjamin Baker of Twinsburg, Ohio, (no. 4,438,584). They use a thick, quicksand kind of glue; the mouse struggles and dies of smothering or a heart attack. That was the novelty, that and the type of glue and the thickness."

Benjamin Baker is modest about his achievement. "To put it in perspective, we didn't invent glue traps per se; exterminators have been using those for years. We developed the first ready-to-use version. Before that, you had to mix the glue and slap it on some cardboard.

"It may not be the perfect trap, but we think it's better than anything else. A lot of people write in pleased because now they've got peace of mind. They can hit the mouse over the head with a hammer or they can let it go—but they know they've got it."

I'll attest to the efficiency of the Bakers' 1979 invention, liberally deployed in the basement laundry room of my apartment building. My objections to what I call the Last Days of Pompeii Trap is chiefly aesthetic. The corpses, their tiny mouths open in pathetic lowercase *o*'s of horror, stand frozen in mid-flight, like those ancient Romans eternally preserved in ash trying to outrun Vesuvius. (Hence the expression "All rodents lead to Rome.") It's egregiously icky. I'm indebted to the inventors for giving me the excuse I needed to send out my laundry. Recently, however, I've been let in on a life-saving secret by a representative of the J.C. Eaton Company, run by Baker *pere et fils.* You can free a still-living mouse from the Stick-Em brand trap by pouring cooking oil on its feet.

Ask a professional, and he'll tell you the best mousetrap is the one that works. "At our convention last year," says Joel Paul, spokesman for the National Pest Control Association, "we had nine companies, each one saying it has the world's greatest mousetrap. Well, it's a good mousetrap if it's neat, quick, and does the job. It's a bad trap if it doesn't."

Among amateurs, Paul says, humane traps are becoming more popular. People don't want to think of themselves as murderers. But they don't want mice, either. Pros have studied the psychology of rodent removal. "We talk about 'the intolerance level.' At the beginning, homeowners are very specific about how the mouse should be taken care of. Once they reach the intolerance level, they'll use any means available. If the Havahart doesn't work, they'll put out five snapping traps.

"Unfortunately, probably because of Mickey and Minnie, mice haven't had bad press. They're made out to be the good guys. But they do cause problems. Mice are very pliable—if they can get their head through an opening, they can get the body through. A mouse has been known to get through cracks the size of a dime. They gnaw through insulation in wires, causing fires. Plus they carry diseases—or rather, disease-causing organisms. Mice are the biggest carriers of Lyme disease. And they're prolific. If two mice, a male and a female, entered your house on January 1, under ideal conditions—all the offspring live and they have enough food and water—by December 31 those two mice will have 2,500 descendants."

Finally, you might ask a mouse. I tried, because I feel sorry for the critter even as I'm frantically hopping up on a chair; it's just trying to eek out a living. But international Spokesrodent Michael J. "Mickey" Mouse refused to comment. "He never does things like that," said his representative at Disney.

About the Emerson thing. His actual gag ran ". . . If a man has good corn, or wood, or boards, or pigs to sell, or can make better chairs or knives, crucibles or church organs than anybody else, you will find a broad hard-beaten road to his house, though it be in the woods." This 1855 passage from his journal was misquoted in an 1899 book called *Borrowings* by Sarah Yule and Mary Keene as "If a man can write a better book, preach a better sermon or make a better mousetrap than his neighbor, though he builds his house in the woods the world will make a beaten path to his door." Emerson may have added the mousetrap shtick in a lecture. In any case, had his original work been correctly quoted, you'd now be reading a droll and informative article on building better crucibles.

A Gripping Yarn

□ ■ □

Fashion designer Jimmy Z pronounced it the love call of the eighties: the rrrrip! of Velcro pieces being peeled apart. The no-fumble fastener has replaced zipper, button, snap, and hook on enough of America's clothing to make its distinctive strip a stirring new sound of seduction—the signal that something's going on because something's coming off.

But suppose you're making war, not love. Today's Army has busted its buttons and promoted Velcro to ranking uniform pocket closure. In combat, that telltale rip could mean R.I.P. if the noise were to reveal a soldier's position. That's why the Army asked for silent Velcro.

"We have, in fact, figured out the source of the noise and informed the Army," says William Kennedy, vice president of research and development for Velcro Group, part of the company that makes the original hook-and-loop fastener. The stuff is woven, usually of nylon, in such a way that thousands of tiny hooks on one side engage thousands of tiny loops on the other. "We've reduced the noise level by over 95 percent," says Kennedy. The mechanics of the new product are hush-hush. "I'd rather not talk about it at this stage of the game. Our patent applications are filed, but they aren't issued yet." But you can know this: the racket is 60 percent the hook's fault and 40 percent the loop's.

The saga of stealth Velcro wasn't the only gripping story to emerge from a visit to the Manchester, New Hampshire, headquarters of Velcro USA. "We're wrongly perceived by many as a gimmick fastener," says vice president of sales Manny Cardinale. "Velcro is fun, but it's a serious product."

Could a gimmick fastener hold together a Pontiac 6000? Could a gimmick fastener keep an airplane wing intact? Could it stick David Letterman to a wall? What was the Jarvik-7 artificial heart—chopped liver? That device's two pumping chambers were secured with Velcro so that if only one malfunctioned, as happened with pioneer patient Barney Clark, doctors could pop it out and replace it without having to remove the other one. "And the space shuttle couldn't have flown without us," Kennedy says.

You get the feeling that the 600 employees here at headquarters think of the space program as a spinoff of Velcro research. "There were 10,000 square inches of tape on each shuttle," adds Leon Kuiawa, market manager of Velcro's government services division. "Everything in the interior—including the astronauts, at times—was held down with Astro Velcro, made with Teflon loops, polyester hooks, and a beta glass backing." And a small patch inside their helmets allows astronauts to scratch their noses.

The word *Velcro* is mentioned, Kennedy points out, as a component in over 5,000 U.S. patents. The company can't, of course, control how others use the product, but it likes to keep track. It's also haunted by Kleenexphobia: dread of going generic. Velcro's patent ran out in 1978; that's meant competition at home and especially in the Far East. Richard Kuhl, vice president of operations, estimates that there are at least two or three dozen hook-and-loop manufacturers in Taiwan alone.

Velcro USA, a wholly owned subsidiary of a Dutch-registered company with holdings in Europe, Asia, and New Zealand, still leads the market, but its name is a sticky subject.

You can call the stuff touch fastener, hook and loop, burr tape, or magic tape, but don't call it Velcro unless you capitalize. A few years ago, Velcro USA considered suing to keep the hairy metal rock band ZZ Top from releasing a record and video called "I Want to Be a Velcro Fly," which sounded not only generic but unwholesome. They settled for insisting that the album cover indicate that Velcro is a registered trademark.

Velcro's a womb-to-tomb kind of fastener, used to strap on fetal monitors and seal body bags. Applications range from the sacred (securing the praying hands of "Special Blessings," a doll from Kenner Parker Toys) to the profane (closing "Cathy's Cuffs," an accessory for those who put stock in bondage). The medical profession would be strapped without hook and loop. (What did they use to close blood pressure cuffs before Velcro? Nails?) The product turns up deployed in unexpected places like the M1A1 tank, where it works like a chain and sprocket to help turn the machine-gun turret, and in nuclear power plants, where a version of the product woven from stainless steel and Nomex, a flame-retardant polymer, holds insulation blankets around radioactive pipes.

Velcro has captured the public's imagination. "We have tremendous customer goodwill," says Kuhl, "although sometimes people accuse us of keeping kids from learning how to tie their shoes." Every week he receives a pile of suggestions for putting hook and loop to work. "Most have already been done, like key rings, pen holders, picture hangers, or wallets," he says, pointing to a thick file of current correspondence next to his Velcro-fastened briefcase. Some include detailed plans; others are hand-scrawled in the heat of inspiration. Most wouldn't be cost-effective for the company, like the removable designer jeans labels that one man proposed. Or there's some glitch, like with the Velcro-soled slippers that kept the elderly from slipping on slick floors but stuck to carpets. The only unsolicited proposal the company has pursued, Kuhl says, was adding a hook-and-loop-closed pocket to beach towels. But no towel company was interested.

Carter Zeleznik, a research associate at the Center for Research in Medical Education and Health Care at Jefferson Medical College in Philadelphia, is undaunted by the company's lack of interest in his globe jigsaw puzzle, on which countries and continents are affixed with Velcro. He'll keep fiddling. "I like the stuff," he says. "I like its natural history. I like the idea that someone got stuck with thistles and learned from it."

The guy who got stuck with thistles was Georges deMestral, a Swiss engineer. In 1948, he returned from a walk in the woods musing over the workings of the cockleburrs that stuck to his socks and his dog. Examining the burrs under a microscope, he discovered that they were composed of hundreds of tiny hooks that latched on to anything even slightly loopy. He figured out a way to duplicate the hook-and-loop configuration in woven nylon, calling the odd product Velcro for *velvet* and *crochet,* French for "hook." Among deMestral's inventions was a Velcro asparagus peeler.

Though the original patent has expired, many parts of the production process are still proprietary. And one nonwoven form of Velcro, a molded plastic hook and loop used in automobile seats, remains under patent. Twenty-four hours a day, five days a week, on clattering high-speed looms, workers weave nylon filament into a loopy fabric that's then dipped in patented goop. On half of the fabric, the loops are snipped to form hooks. "Then the loops are napped to disorient them—a fuzzy surface engages better," Kuhl explains. (That Velcro's loops are more confused than anyone else's seems to make it a superior product. Velcro shoe fasteners, for example, can cycle—open and close, to us laypeople—10,000 times. Kuhl can't vouch for his competitors.) The material is heated to set the nylon's shape-memory, dyed, backed, and cut. Although the nylon product is the company's biggest seller, Velcro's also woven out of Teflon, polyester, and stainless steel.

The nylon version is stronger than people think. Standard Velcro has a shear strength (force applied to parallel pieces pulled in opposing directions) of about 10 to 15 pounds per square inch. "If I use 200 square inches, I've got a ton of force," says Kennedy. "The reason that the David Letterman thing works is that he's got a lot of area."

The host of "Late Night" offered television's most famous, and perhaps only, Salute to Velcro Night in 1986, when he donned coveralls made of hook and made a trampoline-assisted jump onto a wall covered with loop. Letterman reprised the leap for his sixth anniversary show in 1988. Costume designer Susan Hum used 55 yards of two-inch-wide Velcro for Letterman's bright red coveralls. The yellow loop wall couldn't be padded, or Letterman wouldn't have hit with enough force to stick. "David says it kind of hurts," Hum reports.

But it holds. "Some Velcro products have 100 pounds of shear force per square inch," Kennedy says. "Then all I need is 20 square inches to get a ton of holding power." Recently a team of GM and Velcro engineers experimented with using hook and loop in as many ways as possible on a Pontiac 6000—on seats and ceilings, to hold down the battery and fuel lines, to secure the spare tire. It was also tested as a backup attaching device on a front fender. "After the car had been test-driven the equivalent of 50,000 miles, the fender was checked," says Kuhl. "We found that the screws that were supposed to hold it on were never put in or had fallen out. But the fender had budged only a couple of millimeters. Unlike a screw, a hook and loop increases in strength when it's vibrated."

Now the company's negotiating with one of the automotive big three to put together a whole car with Velcro. "And if you can put together a car, why not an airplane?" asks Kennedy.

Members of his thirty-four-person R&D team are working on using Velcro in wing fairings, the one-ton sheets of metal molding that smooth the fuselage to the wing. Usually they're held in place with 14,000 titanium rivets; the idea is to replace some of those with hook and loop.

In the applications laboratory, where the sales staff help customers find innovative uses for hook and loop—"We tell you where to stick it," says assistant Jackie Puglisi—they're working on a project for Uncle Sam: a Velcro-fastened mailbag easier to schlep than one closed with metal-covered strings. A St. Louis physician is experimenting with Velcro as suture material. And for Leon Kuriawa, the word is plastics. "In five to ten years, we'll make an impact in cars like the Corvette, which is made of a plastic composite. Drill a hole for a nut and bolt and you interfere with its internal strength. Bond our product into the structure and you can assemble the car." (One thing you won't see Velcro used for is seatbelts or safety restraints at amusement parks. No matter how strong, it's not more powerful than the potential lawsuits.)

Kennedy and his R&D staff are working on a longer-lived version of Astro Velcro, and on low-cost, disposable Velcro for packaging. The researchers are most interested in custom-engineering performance characteristics. "Suppose we wanted some sort of fail-safe device—something that releases after a certain amount of pressure is applied, like a safety valve. We'd like to engineer the product to release at certain forces. The uses of Velcro are almost unlimited."

It's hard not to be inspired. What frontier is left? Edible Velcro comes to mind. If a material could be spun into a filament, says Kennedy, it could be woven into Velcro. "If there becomes a need for edible Velcro, we'll work on it." If? When was the last time he tried to wrap one of those flimsy pancakes around a pile of mooshoo pork? Don't mooshoo shoot and burrito squirt merit as much attention as loose sneakers?

Not every application has taken hold. The Worm Collar, a cylindrical strip of the loop portion of Velcro that slipped around live bait, the better to snag the teeth of striking bass, seemed like a good idea. But not enough customers bit. Wade T. Fox, Jr., manager of new business development, recently heard about some kindergarteners who played "The Star-Spangled Banner" by opening and closing their Velcro-fastened sneakers, but the product hasn't taken hold as a musical instrument. And the strap-on prosthetic device for impotent men was a flop.

Voodoo Ergonomics

□ ■ □

As you've probably read in the papers or seen on "Entertainment Tonight," Caltech and Princeton astronomers have discovered a light coming from a source they've called the most distant object in the universe. They say it's probably a quasar, the seed of an ancient galaxy, and that its beam has been traveling for fourteen billion years. To get some idea of what an incredibly long time that is, imagine that the total age of the universe is represented by Kareem Abdul Jabbar. The millions of years of human evolution would be merely equivalent to that little plastic thing on the tip of his shoelace. (Which is called an aglet, but that's another story.) Can you imagine piecing together events that occurred that far back? I can't even remember where I was last Thanksgiving.)

A quasar? Yeah, right. (Or yarite—see page 187.) You and I both know that only one thing emits a light strong enough to travel fourteen billion years: the clock display on a videocassette recorder—especially, for some reason, an unset clock. Find the farthest reaches of the universe, my guess is, and you'll find a cosmic home entertainment unit that's been merrily blinking 12:00/12:00/12:00 since the Big Bang caused the primal power surge some fifteen billion years ago. Is God watching *Dirty Dancing*?

The glow from my VCR, kept in the bedroom, could light a major metropolitan area. You can read by it; you can do microsurgery by it; but you'll have a great deal of trouble sleeping by it.

Lately I've been lying awake musing in those tense moments before I get up and toss a T-shirt over the offending appliance: Why do VCRs (and clock radios, to be fair) have such hyperluminescent number displays (function information panels, or FIPs, I would later learn to call them)? Didn't Science figure that some of us had TVs in the bedroom? Who chooses the colors, like the Village-of-the-Damned bluish green on my VCR? Why do so many people allow their VCRs to stammer 12:00 instead of setting the damn things? How can astronomers tell how long light has been traveling through space, and can they tell whether it's traveling first class or coach? Why didn't I cover the flipping FIP before I got into bed, since this happens every night?

I know, I know—FIP glare and related issues aren't nearly as significant to planetary welfare as the greenhouse effect, acid rain, nuclear annihilation, or sweater fuzz. But I've been asking around, and I'm not the only one who's troubled. "Yeah, why *are* they so damn bright?" several people said. And trivial as it may seem, I'm glad I pursued the question, because I ended up learning a lot.

About the power of color, for instance. I began by asking manufacturers how they choose.

Said Lou Lenzi, manager of industrial design for Thompson Electronics, home of RCA and GE VCRs, "Ergonomic principles—the human factor—indicate that something on the red scale is more readable in all lighting conditions, especially dim. We use a combination of red and orange for good ergonomics—orangey for primary information, like the time, and red for secondary information, like the day of the week. Blue is too piercing."

Said Claude Frank, national video project manager of Toshiba, "Red wouldn't do; you can't see it that well. I think blue is easiest on the eyes, but yellow stands out. We use bluish-white on some models and golden yellow on others."

Said Tom Hitzgis, national marketing manager of camcorders for Panasonic, "In general, we find that greenish-blue gives greater readability."

Said ophthalmologist Morton Rosenthal, chief of the retina service at the New York Ear and Eye Infirmary, "There's no evidence showing anything particularly beneficial or harmful about any color of light; none hurts the eye at normal intensity—we're not talking about a flashbulb, here. What seems to matter is whether what you're looking at is in marked contrast to ambient illumination. If you're watching television in a dark room, the glare of the VCR light tends to be tiring, but not harmful. The fact is that some people prefer one color over another."

A guy who's up-front about that is Taiji Izumi, product manager of VCRs for Aiwa. "I tell our Tokyo designers to use bluish-white, based on I like this color. I don't like a noisy color like red or yellow. We use a gentle and sophisticated color, because we make a sophisticated machine."

Sony also chooses whitish-blue for sophistication, according to Aki Amanuma, vice president of the company's Design Center. One determining factor, he says, is what colors are available at the factory. (All VCRs, whatever their label, are manufactured in Asia, mostly in Japan.) Another is what makes the customer feel comfortable, although none of the companies I spoke with pretested devices on consumers. "Light blue is not so aggressive and gives a technical feeling," he adds. An industry insider reveals that a red light display is the cheapest to produce and blue more expensive.

Nothing wilts righteous indignation like a sensible response. I learned that electronics companies are starting to wake up and smell the handwriting on the wall.

"Quite a few calls have come in saying, 'I have this in the

bedroom and it's too darn bright,'" says Toshiba's Frank. "So our newer VCRs dim to half power at 10 P.M. and return to normal brightness at 6 A.M.; it was simple to add that to the memory chip. We started doing that early in 1989; now all our new models have that feature.

"We've only gotten a few consumer complaints saying, 'Why aren't the letters and numbers bigger and brighter?' " Panasonic and Sony also serve TV dimmers.

So the answer to my problem is buy a new VCR, or a thicker T-shirt, and stop whining. And I will, after one little whine before its time: Why was Science so out of touch with what's going on in the American home? "VCR manufacturers, like all other manufacturers, don't take into account how people use their product," says psychologist Donald Norman, chairman of the cognitive sciences department at the University of California at San Diego, and author of *The Design of Everyday Things.* "It's a well-known design problem: How bright should taillights be? How bright should clocks be? Designers tend to overdo to make sure the product is visible. If you're just watching TV or if you're sleeping, the display is distractingly bright. It's on my list."

Norman has collected thousands of examples of everyday objects—faucets, digital watches, stove burners, car doors—designed in such a way that they confound even someone with an engineering degree from MIT, which he has.

(I'd add to his collection my two Trimline phones, each with its clicker-offer placed so that if I tuck the receiver between shrugged shoulder and neck in order to free my hands, I cut off the connection with either my ear or my chin. Yes, I could get a headset, or one of those wedge-shaped neck rests. But I was brought up to believe that the Constitution guarantees all Americans freedom from accidentally terminating telephone conversations with body parts.)

"Designers think they're ordinary people, but none of us is

ordinary in our field of expertise," says Norman, who serves as a consultant to companies like Apple Computers, NCR, and Motorola. Their attitude is, 'I don't have trouble with this; why should they?' " Even one of the companies Norman considers tops, the audio equipment firm Bang and Olufson, isn't perfect. "They're generally excellent: no flashing lights, no extra knobs, everything plain and elegant. But my system consists of four identical units, unmarked, that sit on top of each other. A first-time user can't tell by looking which is the tape deck and which is the CD player. Furthermore, you touch a secret spot to open the CD and change channels—but if you touch it too long, you turn off the machine."

The case of the VCR in the bedroom, Norman says, illustrates one of his simple principles for ergonomic recovery. First of all, products should be tested by real people using the thing the way they're really going to use it. "The best way to test how people will really use a device," Norman says, "is to use the Wizard of Oz technique. You build a fake—as in 'pay no attention to the man behind the curtain'—and ask people to use it as they would if it were the real thing. Suppose you want to test a voice-activated, hand-held electronic notebook, a sort of mini-tape recorder/personal computer, which doesn't exist. You build a little box and put a radio transmitter in it, and you ask people to pretend to use it the way they would the real thing. The designers listen to the requests and instructions people give during their playacting, and you incorporate that information into the real design.

"I worked with an airline on accessing flight information and making reservations via home computers, in plain English, not computer language. No such system existed. To test it, we simply had people type their questions into the computer. Someone at a computer screen in the next room looked up the answers and typed back. The airline learned exactly what people would ask if they thought they were interacting with a computer, how they'd ask it, and what form of answer they'd understand."

Norman declares an object well designed if it offers visual clues to how it works, incorporating what he calls visual mapping. One of his favorite examples is the seat control for a Mercedes-Benz, a button shaped like a seat; you simply push or press the area on the button corresponding to the area you wish to adjust. Norman's own home came with a vertical row of six light switches, each controlling a portion of his oddly shaped living room. Unfortunately, no one in the family could remember which was which. His solution: a switchplate with a layout of the room drawn on it, and each switch located at the corresponding spot.

A good design also takes into account the limits of memory. Norman grumbles over lengthy ID numbers and codes for calling cards and automated teller machine access. (Though I'm proud to say I can, using a series of touch-tone codes, pick up answering machine messages when I'm away, rewind, reset, gather only new messages, and change the outgoing announcement, the processes require so many steps that I actually carry the instruction manual in my purse at all times.)

Finally, a good design anticipates error and makes it impossible. Xerox machines with interior levers that only move one way are a good example. Perhaps the simplest yet most eloquent, Norman says, is the pull-down package shelf in restroom stalls. You can't get out unless you restore the shelf to its original position. (A somewhat similar example is suggested by the colorful and earthy downhome expression "He couldn't put out a fire with a bootful of piss if the directions were written on the heel.")

A sure sign of a badly designed item, Norman says, is that a single control serves multiple and even contradictory functions. Remembering an arbitrary sequence of steps becomes even more difficult. "Take microwave ovens, for instance. If the machine came with a small TV screen guiding you through the steps in English words, that would be much better. Or better still, let the machine read the instructions in the form of a bar code on the food." Another example of this problem, he says, is the setting mechanism for a VCR's clock. "And it's not just Americans who have trouble setting the clocks; I see flashing 12s wherever I travel—in the U.S., Europe, Japan."

So daunting are the clocks (and the machines they live in) that Paul Eskridge, a video and audio consultant, makes a chunk of his living explaining them to people. "Master your VCR!" reads his classified ad in a New York City neighborhood weekly. He makes house calls; just before I talked to him, he'd been giving a woman a refresher course in clock setting. "I point out to people that we can all set clock radios without panicking," Eskridge says. "It's just that the logic of VCR clock setting isn't apparent. I try to make people see the pressing of buttons as logical, not arbitrary."

Mathematician John Paulos, the author of *Innumeracy,* sees the VCR problem as part of a larger national discomfort with high technology. "I don't notice a particular aversion to setting the time on a VCR," he says, "so much as a general feeling of discomfort about machines with numbers. Nobody has trouble with a toaster; there are generally no numbers on a toaster, just the designations *light, medium, dark,* and *burnt.* Maybe VCRs should read *morning, noon, dinner, time to go to bed.*

"I have no further suggestions or comments, except to say I was on the David Letterman show talking about innumeracy and twenty people said they'd be taping me. Only two actually set up the machine correctly."

Electronics firms are on the case. Some companies are moving the VCR clock to a screen display, some plan to eliminate it altogether, and others have arranged for an unset clock to sit there quietly showing 00:00. (Though after a while, a pair of Little Orphan Annies could be as disconcerting as a flashing 12.)

And in November of 1990, Gemstar Development Corporation, a Pasadena, California, company, unveiled a simple electronic apparatus that offers the benefits of marrying someone who understands VCRs without all those tiresome arguments about who stole the covers. VCR Plus+ is a hand-held remote control device that talks to your VCR so you don't have to. First you set it, a fairly simple process that involves punching in a code corresponding to the brand of your recorder and entering the current time and date ("whether or not the time on your VCR is correct," cannily notes a company spokeswoman). After that, you never use the instructions that came with your recorder: When you want to tape something, you punch in a number code published in the television listings of newspapers and magazines; that code turns on the TV and VCR, tunes in the right show, and stops recording when the show is over. So far, Gemstar says, sales of the $59.95 item have been three times greater than expected.

Norman wants to simplify things even further. "The clock is an interesting starting point. In my house, there are about twenty pieces of equipment with digital clocks—the microwave, the stove, the stereo, and so on. It's crazy that every time there's a power failure, I have all these clocks to reset. I'm waiting for someone to invent a single master clock in a single power outlet that sends the time to, and resets, all these pieces. In fact, the whole world could be synchronized by a single clock." (Which is fine by me, as long as everyone doesn't wake up and want to use the bathroom at the same time.)

About that quasar. Called PC 1158+4635—it was between that and "Lefty"—it's been gaining speed since its creation one billion years after the big bang. (How fast is it going? If

I were to paint a 1 and a string of zeros all the way down the Capitol Mall in Washington, D.C., I would be arrested.)

Astronomers know that a light source is receding very fast—and is thus terribly ancient—when its wavelengths increase, shifting to the red end of the spectrum. PC 1158+4635—or is it a VCR?—displays the largest red shift Science has ever seen. The largest red shift I've ever seen was worn by our next-door neighbor, Shirley Bialik—but that's another story.

GENERAL

□ ■ □ ■ □ ■ □ ■ □ ■ □ ■ □ ■ □

The future of our planet may depend on the hybridization of scientific disciplines. In that spirit, *The Bulletin of the Atomic Scientists* might think about adding a swimsuit issue.

□ ■ □

It's a Wise Horse That Doesn't Chew a Hose

□ ■ □

"You can't teach a sleeping dog new tricks," proclaimed an aging relative I'll call Uncle Maxim de Stroy.

"You can't teach an *old* dog new tricks," I corrected. "Really?" he said, leaning forward with interest. "Is that true?"

"I don't know if it's technically correct," I told him. "The point of proverbs is that they're metaphorically useful."

"But are they scientifically sound?" Uncle Maxim wanted to know. "Can the canine learning curve be accurately plotted? Is the early bird really worth two in the bush?" I promised to investigate.

Strange to think that an actual, unsung person must have sat down one day and hammered out a masterpiece of pith like "Don't count your chickens before they hatch." In damp medieval cloisters, men and women probably toiled for weeks on a single saying. "Yo! Brother Acidophilus! Check this out. 'He who hesitates is viscous!' It doesn't sing? Okay. How about 'He who hesitates is fat'? No? Let me work on it." An adage doesn't just happen.

As long ago as 2350 B.C., an Egyptian vizier wrote *The Maxims of Ptahhotpe,* among which is this gem: "Good speech is more hidden than malachite, yet it is found in the possession of women slaves at the millstones." (Probably it was punchier in the original Ancient Egyptian, which read "Beetle, beetle, squiggly line, moon, King Tut–style hat, asp." Proverbs that sound terribly wise in languages you don't know often seem to translate as something like "Better to tap dance with a crustacean than play Yahtzee with the devil.")

Does the early bird indeed catch the worm, and is there some evolutionary advantage to avian punctuality? If you lead a horse to water, can you make him drink? Can you make him do drugs? I decided to ask the experts. And in general, their answers increased my admiration for the anonymous aphorists of the ages.

Apparently, for example, a red sky at night really *is* a sailor's delight; I expected something involving hot tubs and Redi-Whip. But meteorologist David Smith of the U.S. Naval Academy's Department of Oceanography insists it's a red sky, and some sort of honor code prohibits him from lying about what turns on a mariner. "There's wisdom in the adage," he says, "but it's useful only when the sun is behind you. In middle latitudes, weather tends to move from west to east. At night, the sun is to the west. The reddish tint is sunlight being bent by clouds to the east, which means that the weather systems have already passed you, and you can expect a clearing trend. In the morning, the sun is to the east. If you see red, that means light is being refracted by clouds to the west—bad weather is approaching you. 'Red sky at morning, sailors take warning.' " A more helpful saying might be "Red sky at morning, sailors take Dramamine."

Another popular proverb would require only minor alterations to be 100 percent scientifically sound, though its rhythms would be ruined: "You can lead a horse to water, but you probably can't make him drink unless you shove a tube down his nose."

"It's very, very difficult to get a horse to drink if he doesn't want to," says Jack Algeo, head of the animal sciences department at California Polytechnic State University at San Luis Obispo. "I've seen horses hauled from far away refuse to drink water they're not used to. They'll look at it and put their noses in it, but they won't drink."

The tricks used to induce human beverage intake—salty bar snacks, little paper umbrellas, wet T-shirt contests—just aren't going to work on Mr. Ed. "You might add a few drops of flavoring to the water—horses like molasses, or Jell-O flavors like strawberry or raspberry," Algeo suggests. "If you absolutely had to get fluids into a horse—though I've never seen one that dehydrated—you could run a tube through the nasal passage and into the esophagus. If you try to pass a tube orally, he'll chew on it. A horse can chew a hose in half." Waiting's the best ploy, Algeo says. Within twenty-four hours, even the most intransigent neigh-sayers are usually slurping up a storm.

Next I learned the correct thing to mutter when your boss's idiot nephew gets another promotion: "Blood is 10 to 30 percent thicker than water!" According to hematologist Daniel Weisdorf, M.D., of the University of Minnesota, blood's thickness can be determined through a procedure called viscometry, which measures how long a certain amount of blood takes to drip through a thin glass tube. This method can be used to prove the proverb, Weisdorf says. "If you run equal amounts of blood and water through identical tubes, you'll see that the blood takes longer to drip out than the water does."

Geologists I spoke to were adamant about the fact that a rolling stone gathers no moss. "If indeed the particle is rolling, it would be abraded, or worn down, not accreted, or added to," says Karl Koenig of Texas A&M. "Anything accreted to it would be scraped away as soon as it impinged on another surface." But I felt it was important to get the moss's side of the story. (Right—I didn't want to take anything for granite.) So I consulted William Buck, curator of bryophytes at the New York Botanical Gardens. "I don't see how a rock can pick up anything," he says, "be it moss or a date for Saturday night." And that's the last word on moss transit.

The whole bird/worm thing doesn't tie up quite so neatly. "Those birds, such as robins, that eat worms are more likely to catch them in the morning, because worms come to the surface when the ground is moist and cool and disappear in the heat of the day," says Mike Cunningham, associate curator of birds at the Los Angeles Zoo. But because their metabolic rate is so high, Cunningham says, most birds feed morning and evening. So presumably a late bird could also get a worm? "A worm-eating bird looks for worms whenever it can get them."

If you're more interested in catching flies than worms, don't be so sure you'll nab more with honey than with vinegar. "There happens to be a kind of fly called the vinegar fly," says entomologist Al Grigarick of the University of California at Davis, "which is attracted to fermenting odors. If you want to attract the greatest variety of flies, however, honey is probably your best bet. But neither honey nor vinegar will *catch* any flies unless it's used as bait in a trap; the flies won't land in either substance, but beside it." The best way to use sweets as bait, Grigarick says, is to soak string in sugar solution and insecticide. "Houseflies seem to be attracted to stringlike objects," he explains. "They have sensory receptors that allow them to taste with their feet as well as their mouthparts, and they pick up the insecticide." I think the moral here is keep your big mouthparts shut and don't eat with your feet.

Unless you're feeding a cold. Or is that starving a cold? I'd heard somewhere that the expression "Feed a cold, starve a fever" originated perhaps 1,000 years ago and employed the Old English word *steorfan,* "to die." Thus its true meaning, according to this theory, was "Feed a cold *and you will die of a fever.*" Close, but no stogie. I ran the theory by George H. Brown, professor of English and classics at Stanford University. *Steorfan*—and subsequently the Middle English word *sterven,* did once mean "to die," he confirmed. But by the nineteenth century, when the expression "Feed a cold, starve a fever" first appears in literature (and colds and fevers were epidemic in England), *starve* had been in use in its modern sense for at least 300 years. Nineteenth-century physicians shared with common folk the belief that a cold, thought to be caused by a chill, could be cured by heating up the body, with food as fuel. A fever, on the other hand, required cooling, achieved by fasting. This belief was probably based on the writings of Hippocrates: "As the soil is to trees, so the stomach is to animals. It nourishes, it warms, it cools; as it empties it cools, as it fills it warms."

An etymologically satisfying explanation, but medically irrelevant. "A cold is caused by a virus; it's self-limiting," says internist Joel Curtis, M.D., of Beth Israel Medical Center in New York. What or how much you eat has no bearing on the length or severity of the illness.

But Ogden Nash, purveyor of pedigreed doggerel, was right on the money when he maintained in his poem "Reflections on Ice-Breaking" that "Candy/is dandy/But liquor/is quicker." "Alcohol begins its absorption directly across the stomach," says Curtis Wright, a researcher in behavioral pharmacology at Johns Hopkins School of Medicine. "Blood alcohol begins to rise almost immediately after ingestion. With the exception of certain forms of glucose scientifically prepared to be absorbed in the mouth, sugar can't be absorbed until it moves from the stomach to the small intestine. Most candy contains sucrose, a disaccharide, which has to be split into two sugars, glucose and fructose, before it can be absorbed."

And that's why alcohol is so much more efficacious than Milk Duds as a social lubricant.

Candy, like any carbohydrate, creates a feeling of well-being by triggering an insulin surge, that ultimately increases the level of endorphins, the brain's feel-good chemical. (See "Lifestyles of the Rich and Creamy.") How does alcohol cast its spell? "You could stir up a lively debate among scientists as to what exactly is involved in intoxication," Wright says. "Some people believe that alcohol actually produces a narcoticlike substance in the brain, but that theory hasn't held up too well recently. There's also a hypothesis that alcohol produces a swelling of all the membranes of the body. [Hence the expression "a swell party."] That swelling distorts certain chemical receptors in the brain that are very closely related to the barbiturates and diazapines—Valiumlike drugs that are generally thought to be euphorigenic. These distorted chemical receptors then fire abnormally, as if a drug were there."

One intriguing, highly specialized subset of folk wisdom is parental guidance: "Your face is going to freeze in that horrible expression!" Yeah, right. (Or yarite. See page 187) Where do moms and dads pick this stuff up? It must be genetic, or maybe hormonal, because once we reproduce, otherwise sane and admirable people such as ourselves are possessed by devils—Ma and Pazuzu—who make us say things like "Will you be happy when you've broken it?" and "Sure it's fun, until someone loses an eye." Speaking of which, a favorite in the parent's admonition derby is "Don't read in that light, you'll ruin your eyes." Wrong. "People falsely equate illumination with vision," says ophthalmologist Hampton Roy of the University of Arkansas Medical Center. "Sometimes people over forty who suffer from presbyopia [an inability of the lens to change shape in order to focus up close—in other words, aging eyes] can see better when their pupils contract in brighter light. Perhaps because some mothers see better in brighter light, they think their children will see better in brighter light. But as long as you can see the page, you're fine." In fact, intense ultraviolet light, both natural and artificial, is actually bad for your eyes. Sidney Lerman, a professor of ophthalmology at Emory University who's been studying the long-term effects of ultraviolet light, says that it causes at least 10 percent of the cataracts in this country. He hastens to add that testing can reveal vulnerable individuals and that with proper protection, such cataracts can be prevented.

Maybe when your mother told you not to read in that light, you'd whine, "But Johnny's mother lets him." That really fried your mom's hide, didn't it? "And if Johnny told you to jump off a cliff," she'd ask acidly, "would you do it?"

The scientific answer to your mother's question is "Probably not." According to child psychiatrist Charles Binger of the University of California, San Francisco, although acceptance by peers is terribly important to young people and struggling to separate from parents is a major psychological task of adolescence, most kids won't abandon family values in the face of peer pressure. "Studies of normal kids show that contrary to popular belief, they go through adolescence without being particularly rebellious. The better the home life, the less likely the child is to flee to peers, and the more likely he or she is to choose peers wisely."

Finally, I wanted to learn the scientific truth behind a popular rhetorical device used when the reply to a question is obviously affirmative. We might call it the theological-ursine-scatalogical-sylvan interrogative series: Is the Pope Catholic? Does a bear—let's say sit—in the woods? The first statement is indisputable. But what's the straight poop on the second? "If you want to get picky," says Penny Kalk, collections manager of the mammal department at the Bronx Zoo, "it would depend upon the bear." Polar bears, she points out, sometimes drift for days on ice floes; they may answer the call of nature there or in the water. "But the American black bear, the animal most people have in mind when they think 'bear,' lives in the woods; that's where they eat, and what goes in must come out."

I still have much to investigate before I report back to Uncle Maxim. Are scientists absolutely positive, for example, that not the tiniest quantum of benefit can be wrung from crying over spilled milk? In order for a wind to be ill, does it have to blow absolutely nobody any good, or can it blow the majority of the population no good, a few some so-so tidings, and a couple of people dead leaves and scraps of old newspaper?

This project will take a while. Do I think it'll be worth the trouble? Is the Pope Catholic? Is good speech more hidden than malachite?

The Comma Before The Storm

□ ■ □

The experts say I'm going to have trouble slipping the new punctuation marks I've invented—the irol, the yarite, the stomapod, and the goober—into modern English usage.

They say I don't know enough about the venerable history of punctuation, that I haven't fully contemplated its rhetorical and grammatical raisons d'être and that, like most of us, I've failed to acknowledge punctuation's surprisingly lofty role in the evolution of scientific thought. They say it's irresponsible to bring new punctuation into an increasingly unstable and ungrammatical world where two of our most beloved and versatile marks are in grave danger of extinction. They say that "goober" is a stupid name.

My first responses were "Kiss my *" and "What do Raisinettes have to do with punctuation?" The punctuational world needs shaking up, I sez; not much has happened since quotation marks hit the scene at the end of the seventeenth century.

(Except, of course, for the interrobang. In the early 1960s, the late Martin K. Spektor, a New York advertising executive, decided our language lacked the proper garnish to fully complement rhetorical questions framed in such a way that a question mark is technically called for but an exclamation point—a "bang" in typesetters' slang—would be more emotionally satisfying: Isn't that something‽ Why don't you go defenestrate yourself‽ You're what‽ So this is how the world ends, not with a whimper but with an interrobang‽

(The actual shape was created in 1967 by Dick Isbell, a noted designer of typefaces who's now a senior graphics specialist at General Motors, where he dreams up the lettering for car names. "I put an interabang—that's how I spelled it—in at least a hundred alphabets I designed," he recalls. "But it never caught on as a punctuation mark."

(The story of the interrobang can be interpreted as cautionary: The English language hasn't had a new punctuation mark in two centuries because it doesn't need one. Or it can be considered encouraging: Our planet went fifteen billion years without a single "kid/adult switch bodies" movies until the 1977 nonclassic *Freaky Friday.* A decade later came a string of undistinguished spawn: *Like Father, Like Son, Vice Versa, 18 Again!* then, finally—ta da—*Big,* a critical and box office smash. If the interrobang is the uncelebrated Barbara Harris vehicle of punctuation, perhaps the goober will be its *Big.*)

The earliest recognizable ancestor of the punctuation mark was probably invented in about 260 B.C. by an Alexandrian Greek librarian, Aristophanes of Byzantium, also known as Aristophanes "The Librarian Not the Playwright" of Byzantium.

"Punctuation began as an attempt to transform the written into the oral, to restore all that had been abstracted away: rhythm, breath, emotion," says Robert Logan, a University of Toronto physicist, author of *The Alphabet Effect,* and the logical person on whom to test the goober, irol, et al. Logan's special interest is the history of science, specifically the way the invention of the alphabet has influenced thought patterns and social institutions, leading directly, he believes, to the development of deductive reasoning and thus the scientific method.

The Sumerians, Logan explains, invented pictographic writing in 3100 B.C.; 1,500 years later the Canaanites, a Semitic tribe of the Middle East, refined that notion into the first phonetic alphabet, ancestor of all the world's alphabets from Russian to Swahili. Not all writing systems have been alphabetic, he points out. Some, like Babylonian, employed syllaberies, a set of symbols for each syllable rather than each phoneme, or individual sound. Others, like Chinese, have remained pictographic; each basic symbol—there are over a thousand—stands for an entire spoken word, not an abstract sound. (Early attempts at a Chinese version of alphabet soup were abandoned when it became clear that not bowls but vats would be required.)

Alphabetic writing, Logan contends, contains subliminal lessons that pictographic writing doesn't, among them: (a) the conversion of auditory signals into abstract visual signs, (b) analysis, (c) classification, and (d) ordering. Pictographic systems, he says, foster a different kind of thinking, more concrete and less abstract. That's why, he believes, theoretical science developed in the West and technology and its practical application in the East. Fascinating, but what about my goober? Logan says he's getting to it.

"I'm not sure the punctuation effect was as strong as the alphabet effect, but it did contribute to making written language uniform and consistent. And in that way, it contributed to the rise in abstract and rational thought and thus to modern science. It's interesting that Chinese has been traditionally unpunctuated."

Punctuation didn't exactly burst into the writing scene, says Ellen Lupton of New York's Cooper Union for the Advancement of Science and Art, who researched and mounted an entire gallery exhibit devoted to punctuation. The Greeks improved on early Semitic alphabets by adding vowels. (The first practical result of this innovation was, of course, the IOU.) "But early Greek inscriptions and, later, manuscripts had no spaces between words and no punctuation, and were written in all capital letters of uniform height." LIKETHISTOUGHGOINGHUH Not only that, they employed a form of writing called boustrephedon, meaning "as the ox plows." The first line of a work might run left to right,

,htgir ot tfel txen eht ,tfel ot thgir txen eht

and so on. In a few inscriptions, phrases are separated by a vertical row of two or three dots. Aristotle mentions only one form of punctuation, the paragraphos, a horizontal line drawn under a word introducing a new topic.

Anyway, Aristophanes got bored telling patrons not to rustle the papyrus and invented a dot system of punctuation based on Greek rhetorical theory, which divided a written work—meant to be read aloud—into rhythmic sections of different lengths. He marked the end of a short section, known as a comma, with a point or dot placed after the middle of the last letter. A longer section, called a colon, was marked by a low dot and the longest section, or periodos, by a high dot. (This evolved, of course, into the period, which today is roughly the size of the point at the end of this sentence.)

This new system ought to have launched a great epoch known as the Period Period, but it failed to catch on; dot's all she wrote until the fourth century A.D., when a Roman grammarian called Donatus (I'd like to think his first name was Duncan) revived Aristophanes' scheme. Even so, most Roman works remained fairly innocent of punctuation, except perhaps for points (puncti) between the words of an inscription, or a mark within the text called a capitulum, much like our modern paragraph sign, which signaled a change of thought.

During the Middle Ages,[1] when manuscripts were literally handwritten,[2] mostly by monks, Aristophanes' system metastasized. "Punctuation, like spelling, was wildly idiosyncratic," Logan says. The marks and their purposes differed from country to country, city to city, and even monastery to monastery. By the end of the eighth century, European scribes employed some version of the period, comma, colon, semicolon, question mark (originally horizontal), and slash or virgule, used like a comma. Sections were sometimes distinguished by a darling little leaflike thing called a hedera, Greek for "ivy." Scribes might use asterisks and daggers to mark footnotes, or simply as decorations to finish out a short line.[3] Punctuation remained keyed to speech, indicating rhythm, inflection, and emphasis, because nearly all reading was done aloud, even when people read to themselves.

The invention of printing brought uniform punctuation and a transformed world, though not necessarily in that order.[4] The sixteenth-century Venetian scholar and printer Aldus Manutius published the first guide to improved and codified punctuation; *he—or his thrifty typesetter—also invented italic type,* designed to take up less space and thus save money. By

1. As time marches on, won't the designation "Middle Ages" eventually have to be changed to something like "Pretty Near the Beginning Ages"?

2. Scribes used pens, of course, not just their hands; I meant that to be an etymological literally: manuscript, from the Latin *manus,* meaning "hand," and *script,* Latin for what every waiter and waitress in Los Angeles is working on.

3. Punctuation wasn't used to indicate swearing until the early twentieth century, and if you don't believe me you can $#%&*! yourself.

4. The mass printing of books meant a postmedieval revival of learning, the spread of literacy, and the advent of silent reading, which contributed to the rise of individualism, capitalism, and, ultimately, a McDonald's on Red Square.

the end of the seventeenth century, quotation marks, dashes, and parentheses had appeared (an especially mysterious phenomenon when there was no printed page nearby). And by the end of the eighteenth century, the rules of modern punctuation were more or less established, though there have been fads and trends like the Elizabethan orgy of overpunctuation and the punctuational minimalism, or perhaps lethargy, of our own age.

This won't be a usage lesson; grammarians are a tough lot and I don't want to wander onto their turf. (As a testament to their seriousness, consider the dying words of French grammarian Dominique Bouhours (1628–1702): "I am about to—or I am going to—die; either expression is used.") Suffice it to say that since the invention of the printing press made reading primarily a visual and not an oral pleasure, punctuation no longer serves its original purpose as a rhetorical device indicating pause and inflection. The modern reader isn't in much danger of asphyxiating for want of a comma. Rather, we need punctuation for grammatical purposes, to clarify the precise meaning of a clause or phrase.

Take an unpunctuated sentence used as an example in an early 1930s article in the *American Bar Association Journal* about how misplaced punctuation can lead to confusion (and in some cases, to litigation):

> Woman without her man is a savage

Only punctuation can tell us exactly what the author had in mind here. Some might leave the sentence unadorned but for a period. Others might add commas: "Woman, without her man, is a savage." And still others might prefer "Woman: without her, man is a savage."[5]

5. The comma is also useful in what might be termed punc rock, viz. the lyrical break in "Handyman" ("Comma comma comma comma com com")

Yes, conventional punctuation has been doing a fine job. But I believe that life in the waning days of the twentieth century has given rise to situations not adequately covered. That's why I'm proposing some new marks:

The irol and yarite will, I hope, answer a question that has been around since before 1785, when the British scholar Joseph Robertson raised it in "An Essay on Punctuation":

> What pause is proper after an ironical expression? In answer to this enquiry, it must be observed that there are two forts [*sic*] of irony, the grave and the exclamatory. The former may be terminated by a period; the latter, by a note of exclamation. . . . Some writers have asserted, but, I believe, without foundation, that the Germans mark an ironical expression by inverting the note of exclamation thus: What an admirable poet¡ We have no such mark of distinction; because perhaps it may be supposed that the character of the person commended, the air of contempt, which appears in the writer, and the extravagance of the compliment, will sufficiently discover the irony, without any particular notation.

Well, maybe. But I favor the German notion, though "inverted exclamation point expressing irony" seems an unwieldy name. I propose irol, pronounced EYE-roll and indicating that the statement it follows would be spoken with the eyes rolling heavenward: "*I Dismember Mama* is my favorite movie¡"[6]

A possible substitute for the irol is the yarite, a tilted tilde meant to resemble a smirk and denote an unspoken "Yeah, right":

---

and "Breaking Up Is Hard to Do" ("Comma comma down dooby do down down). The dooby, of course, fell out of use in the twelfth century.

6. This mark could also be called the volvo, a homonymously sound choice: in Latin, *volvo* means "I roll."

As I understand it, my esteemed colleague from Illinois believes his bill will erase the deficit by the year 2000⸮

Vice President Quayle now knows that people in Latin American countries don't speak Latin and won't make that mistake again⸮

I'd be happy to redo that report, boss⸮

The goober is the philosophical and philological cousin of the interrobang. It looks like an Rx sign with a *u* instead of an *x*; the *u* has an extra-long tail and from it grows a little stylized peanut. It means "Are you nuts?"

You want this by tomorrow ℞

Sentences requiring the stomapod most often begin "I probably shouldn't be saying this, but. . . ." Right. You *shouldn't* be saying this. But you do. And then you realize you've put your foot (Latin, *pod*) in your mouth (Latin, *stoma*). I suppose this would look like a stylized foot in a mouth, but I don't know how to draw one.

When I ran these marks past the experts, they responded with polite interest and just the faintest suggestion that even as we spoke they were pressing a silent alarm underneath their desks that would summon a kindly but armed person who would escort me out of Academia.

Said Ben Blount of the University of Georgia, an expert in anthropological linguistics, "It's harder to introduce punctuation and usage than it is to get rid of it—not as much agreement or motivation is required. There's some discussion these days about eliminating the semicolon as unnecessary. In France and Spain, a national academy rules on such questions. Early in American history there was a move to set up an American academy, but it was rejected; it represented the kind of authority the Revolutionary War had been intended to overthrow."

If there were some sort of official academy of language in this country, he seemed to be suggesting, it could rule on the irol, yarite, etc., but he was glad there wasn't. And he really wasn't crazy about the name *goober.*

Linguist Richard Ohmann of Wesleyan University gently cautioned that I might need a century or so to make the goober fly. "Punctuation is not dynamic. It's much more stable than word usage, a matter of printers' conventions which stand still. Like spelling, punctuation has been 99 percent constant for 200 years."

Logan was more sanguine about the dynamism of punctuation. "I predict that the forward slash and the backward slash, used on computers, will trickle down to literary use, simply because they're there. I'm not sure how they'll be used. Maybe a forward slash will indicate a fragment and a back slash a non sequitur; maybe we'll use back slashes to indicate reminiscences and forward slashes to speculate about the future."

I sent my inventions to a large and influential publishing house. They wrote back that the marks were terrific and the check was in the mailʔ

Cheese It—The Cops! or The Dairy Case

□ ■ □

The woman on the phone had a voice that could melt the wax off a gouda. U.S. Customs, she inflamed, had intercepted a load of imported Cheshire that looked suspiciously like cheddar sneaking into the country under a fancy alias. Probably trying to dodge import quotas. Could Mike give them a positive ID?

Mike Tunick of the U.S. Department of Agriculture, Mike Tunick, chemist, Mike Tunick, *cheese detective*—took the gig.

Tunick gets the kind of case that won't yield to the probing of your average shamus, who doesn't know jack cheese about dairy products. Regular police procedures don't work either. You know what they say: You can grill cheese all night and it still won't talk. That's where Tunick, thirty-five, comes in. Mild-mannered but no Milquetoast, he knows how to pressure a tough customer. He does it with a texture profile analyzer—a machine that measures stress tolerance by mimicking the chewing process—instead of with threats. None of that "Yo, curdball, keep it up and you'll be huggin' macaroni" stuff. And the only heat Tunick packs is the differential-scanning calorimeter he uses to measure the melting points of various suspects. Often that's the only way to unmask an adulterated mozzarella, or maybe a "butter" that's known to its close personal friends as Oleo.

His turf, the USDA's Eastern Regional Research Center, is the venerable fifty-year-old lab where instant mashed potatoes and concentrated fruit juice were invented. Tunick's attached to the Milk Components Utilization Research Unit. The lab is a mile or so outside Philadelphia, where Tunick grew up. As a boy he showed no early signs of a calling. A healthy appetite for cheese steaks, sure, but he didn't sit around reading Havarti Boys mysteries. In 1973, when most other college kids were doing pot cheese, Tunick started at the center as a student intern; he was hired full-time in 1977.

Tunick introduces his boss, chemist Virginia Holsinger, one of the researchers who devised a milk for the lactose-intolerant. She heads the Milk Components Utilization Research Unit's twenty-person team. "We do the kind of longer-term, higher-risk research that universities can't afford to do," she says. Recently the unit took a shot at keeping good cheeses from going bad, and it paid off. "Warehoused cheddar earmarked for special federal projects like the school lunch program was ripening faster than the government could use it," she explains. "Three years of experimentation and two more of shelf-life studies proved that it's possible to freeze a large volume of cheddar, then thaw and process it without any significant damage."

The units launched a major new project, making low-fat, low-sodium cheese that doesn't feel and taste like a rubber duck. "We're starting with mozzarella, trying to get a handle on how variations at the molecular level will affect texture and melting.

"The detective work—we've done milk and butter, not just cheese—is only a sideline."

But don't the gumshoe and the research scientist walk the same mean streets? What is it but sleuthing when you're hunting down new uses for discarded beef fat, Tunick's first project at the center? (He and colleagues used the fat as a cheaper alternative to cocoa butter in some chocolate products.) Aren't you Science's own hawkshaw when you're trying to find a clever disposal method for tannery wastes, as Tunick did when he worked for the center's Hides, Leather, and Wool Research Unit in 1977? ("We came up with a way for microorganisms to chew up waste and create methane gas you could use to heat the tannery. We made some grand messes.")

But let's face it. Research most resembles detection when a crime has been committed. That's why the mozzarella caper, Tunick's most successful case, was such a kick.

"The basic problem," he says, "was that this company, Dakota Cheese, was adding calcium caseinate powder—a cheap milk by-product—to extend the yield of its mozzarella by about 25 percent. Then they sold two million pounds of it to the National School Lunch Program as natural, not imitation, mozzarella."

USDA inspectors got suspicious when they saw bags of calcium caseinate sitting around, about as inconspicuous as a scorpion on a wheel of brie. But they needed proof that someone was cutting the cheese.

"My work on the differential scanning calorimeter is what did it. I started out looking at water content differences between real and imitation mozzarella, freezing the cheese and melting out the water. That didn't show me a lot." But he happened to notice something about the melting of the fat in the cheese. "It turns out that calcium caseinate works as an emulsifier. So when we refrigerated the cheese, the fat stayed liquid instead of solidifying like the fat in natural mozzarella would. The more calcium caseinate you had in there, the more it would stay liquid. So I would just pull the stuff out of the refrigerator, stick it in the instrument, and heat it up. The smaller the melting peak was, the more caseinate was in there.

"Then through the electron microscope we saw that the fat looked different in the two kinds of cheese. Globules in natural mozzarella are small. But the more calcium caseinate that's been added, the more the fat globules tend to coalesce and form giant globules. We got electron micrographs of that."

And there his job ended. "We're not the enforcers," Tunick explains. His reports can't be used as evidence, and he and colleagues can't even testify, because theirs is a research rather than a regulatory body. Whatever evidence he noses out has to be corroborated by a certified lab. Someone else tells the court his good news.

Until Tunick's mozzarella report showed up, Dakota Cheese intended to plead not guilty. But there's nothing like an attractive album of electron micrographs to change a guy's mind. Dakota and its president went for the civil penalties and ended up convicted on seventeen counts of fraud. "The fines amounted to $515,000," Tunick says. "The government figures they lost three quarters of a million dollars on the substitution of imitation for natural. I was told that kind of recovery—70 percent—is pretty good."

Adds Tunick, "I'm also involved in this cheddar/Cheshire controversy, which has yet to be resolved. Cheshire is sort of like cheddar, only drier. When this cheese came into the country, it didn't look like Cheshire. It looked too yellow to be anything. At first we couldn't tell with a hundred-percent certainty what it was, because to elude detection they had ground the cheese and reprocessed it. The differential-scanning calorimeter didn't tell us anything. So we turned to rheology, the study of flow and deformation of matter. Basically we found that the textural properties, the viscosity, of the two were different enough so we could tell that what was labeled Cheshire really wasn't.

"There's no quota on importing Cheshire into this country because we don't make it here, but there is a quota on foreign cheddar. Now there's a U.S. importer in trouble."

Tunick heads for a local joint when the lunching lamp is

lit—Edam if you got 'em. But they don't, so Tunick orders the French onion soup topped with melted mozzarella. That way he can demonstrate some simple flexibility tests—pulling strings of cheese ceilingward—while he relaxes. His main course is a Cheshire (authentic) burger.

"The apocryphal story about the origin of cheese is that a guy was riding a camel through the desert carrying milk in a pouch made from a cow's stomach," Tunick says. "Rennin, an enzyme found in a cow's stomach, caused the milk to curdle and harden. And this brave soul ate it anyway." Tunick thinks the oldest physical evidence of cheese, besides something green and hairy in his refrigerator, is residue in a Sumerian artifact from around 4,000 B.C.

Tunick is a man with dreams but no illusions: Give it any pretty name you want, he says, but you can't dodge the fact that cheesemaking is the art of controlled spoilage. To make a ripened cheese, he explains, you start with any mammalian milk. Even weasel, if you're that sort. Ferment it with bacteria and turn it into curd by adding an enzyme, usually rennin. Then press out the liquid—no whey, José—salt it, and wait.

A hard cheese does hard time—three to twelve months at 35 to 50 degrees Fahrenheit. Cheddar and Swiss are made this way. The holes in Swiss—known as eyes—are carbon dioxide bubbles emitted by the bacteria busy breaking down fat and protein, Tunick explains.

Soft and semisoft cheeses aren't much different from most of humanity: they either rot on the surface or from within. Camembert, Brie, and Limburger are surface ripeners. Their bacteria sometimes raise a stench not unlike that caused by bacteria that hang around the warm, moist regions of the human body. They say the French surrealist poet Leon-Paul Fargue had the guts to call Camembert *les pieds de Dieu*—the feet of God.

Blue-veined cheese ripens from the middle. In Roquefort, for example, the blue mold *Penicillium roqueforti* is added at the milk or curd stage; it stays dormant until activated by carbon dioxide in natural or artificially created cracks in the pressed cheese. To be called Roquefort, the cheese has to spend three to four months in a limestone cave in the village of Roquefort in the south of France. Without the cave, a Roquefortlike cheese is merely *bleu.*

"A dessert without cheese is like a beautiful woman with only one eye," said Brillat-Savarin. Nevertheless, over dessert Tunick orders a chocolate, not cheese, crepe, and tells me about another case. In late September, Customs called the lab again. Something had entered the country labeled hard cheese, but officials were suspicious. "I came into the lobby and a guy handed me a 45-pound block. By this time we'd gotten the procedure pretty much down to a science, pardon the expression. It took ten people to analyze the mozzarella. This time it took only me and my colleague, Jim Shieh.

"It was pretty simple. We tested it on the viscometer, the calorimeter, and the texture analyzer. It didn't meet the moisture and fat standards of identity for Parmesan, provolone, Romano, or even hard cheese—there's a standard of identity just for generic hard cheese. Parmesan can be no more than 32 percent moisture, and provolone can be no more than 45 percent. This was a little under 45 percent, so it fit the scheme for provolone, but it didn't meet the provolone fat standard. It didn't meet the standards for anything—which is what Customs wanted to know. I don't think anyone's criminally in trouble; they're just going to have to label it better."

Tunick doesn't consider his profession dangerous, though he knows he could always end up with a fondue fork in the back. "If the international organized cheese crime cartel goes after anybody, it's probably not gonna be me. But I kidded with my wife during the mozzarella case that some night I was going to get into bed and find a severed headcheese hidden in the sheets."

The work can be tedious and trying. Tunick doesn't feel quite like Job, who griped to the Lord, "Hast thou not poured me out as milk and curdled me like cheese?" But some days he looks around the lab and thinks, I belong here like chocolate sprinkles on Gorgonzola. He's a scientist, though, a man with a mission. The way he sees it, he and his colleagues have kept schoolchildren from the brink of malnutrition, improved foreign relations, and saved the taxpayers tons of money and heart trouble. Tunick doesn't get all sickening about it. How does that song go? "I try to say what I feel, but the curds get in the whey." Says Tunick simply, "When you get a conviction and a published paper out of a case, that makes it all worthwhile."

Tunick doesn't worry much about the fates of the cheese cheats he brings to justice. Why brood over a feta compli? "*C'est la guerre,*" he says, shrugging. "*Quelle fromage.*"

Under the Volcano

□ ■ □

Did you know that, according to the National Science Teachers Association, the standard high school gym can accommodate about 200 science fair projects? Did you know that 199 of them are likely to be volcanoes? Did you know that out of 89 Nobel Prize and Medal of Science winners recently polled on the subject, only two had taken first place at a science fair?

For the 1990 book *Science and Math Events: Connecting and Competing,* editor Deborah C. Fort asked a platoon of luminaries to gauge the value of science fairs and other competitions. Fifteen percent declared them irrelevant and even damaging; 35 percent said competitions are generally beneficial. Half were undecided about the value of contests, some because science fairs didn't exist when they were students.

No one knows for certain when the science fair emerged as an event, says Fort, who's seeking a grant to research the subject. "No written history exists. There's a controversy, though not a passionate one, about whether it began with the Vegetable Growers Association of America's growing, judging, and identification contest, first held in 1935; the Westinghouse Science Search, launched in 1942; or the International Science and Engineering Fair, opened for the first time in 1950."

I tend to agree with the experts who believe that soon after the earth cooled, the cosmic soup spawned cosmic salad, from which sprang protoscience fairs, tiny one-celled creatures that eventually crawled onto shore and rented cabanas. Sometime between the Late Cretaceous and Early Bodaceous periods, they began traveling in packs. Soon 200 of them fit into a standard high school gym.

Others insist the first event was a happy mistake, like Charles Goodyear's spilling rubber on the stove and discovering vulcanization. They theorize that science fairs occurred when a researcher trying to synthesize a Singles Weekend in the Catskills accidentally added too many dentists. He shoved the botched project into a storage shed and forgot about it until weeks later when he was hunting for his Garden Weasel and discovered the experiment had mutated into an affair involving cafeteria tables dressed in skirts.

For this reporter, Fort's findings—more about them in a moment—opened the floodgates of memory. (You might want to run for higher ground.)

It is long ago, but not *that* long ago. At Orchard Dale Elementary School, in eastern Los Angeles County, a little girl is feverishly dyeing sand different colors and patting it into strata in a cracked aquarium. She's creating a cutaway view of planet Earth. But the sand is still wet on the day of the competition, and the red crust and yellow mantle sort of run together, causing one judge to quip to the little girl that this was the first time anyone had ever entered lasagna in a science fair. The little girl doesn't know what a quip is; she only knows she hurts, especially since her nemesis, Lylette Stanley, Miss Priss herself, has made a model of the eyeball that looks like it came out of the National Eyeball Museum or something.

The little girl smiles through her tears, but she swears to herself that when she grows up, she'll get a job making fun of science in books and magazines. She never enters a science fair again. And today that little girl is, of course, Christie Brinkley.

Listen, I probably have a bad attitude because I never got over my disappointment at discovering the true nature of science fairs. Maybe it's just me, but when I hear "fair," I think "rides and games!" Wasn't it a major letdown the first time you shuffled into that gym and found not even a ferrous wheel or a benzene ringtoss?

Here's one good thing; I was in the forefront of the anti–Big Science backlash. The supercollider is budgeted at $8 billion. My project came in at $7,999,999,998 less.

You know, even if it hadn't looked like a tank of puke, my project still wouldn't have fulfilled what ought to be the true spirit of the science fair.

"Instead of doing science, kids do art projects," says Gerald Garner, director of the California State Science Fair. "They make charts, posters, models. A kid should understand what a hypothesis is, collect data, metrify everything, be able to make some graphs, and come to some conclusions that have something to do with the hypothesis.

"You don't have to spend money to have a good science project. You can count blades of grass in a sunny spot and a shady spot, do some statistical comparisons, and that's good science. Sure, kids are fascinated by making a working volcano. It's not bad for them, but it gives them a misconception about what it means to do science."

Wise words. But what caught my attention was Garner's spontaneous mention of volcanoes. He'd tapped into precisely the trend that had emerged—no, erupted—from an informal poll I'd taken of just plain folks, friends, and strangers. I'd ask about their science fair careers, and suddenly it was the Last Days of Pompeii.

(Note: Some of these people felt completely comfortable using their names, even when publicly discussing as intimate a subject as science; others, generally those whose earliest encounters were painful or embarrassing, insisted on anonymity. I checked around to see how other publications handle this, and the locution of choice seems to be "Tina [not her real name], a computer programmer from Denver. . . . So be it.)

Tina (not her real name), a computer programmer from Denver who grew up in Berwyn, Pennsylvania, told me that a volcano kept her from ever entering a science fair. "In fifth grade, a boy in my neighborhood, Richard Martin (his real name), made a volcano that actually erupted. He was keeping it in his garage until the science fair. I went over to his house and begged him to erupt his volcano for me, but he wouldn't. He said he was saving it for the teacher. He was my first—but not my last—withholding man. Maybe he had a headache. In any case, I was so disappointed I lost interest in science fairs forever."

A volcano also turned off Joe Millett, an assistant at a New York literary agency. "In fourth grade, at the Buckley School in Roslyn, New York, some kid made a papier-maché volcano, very elaborately rigged. Underneath he had a blob of dry ice in fine wire mesh and an electric fan, to create the illusion of smoke. That's all it would do—no lava. I remember that he had big rubber gloves on, and he was lifting off the top of the volcano to put the dry ice in the mesh bag when a third grader couldn't resist licking the ice. Naturally, his tongue stuck. The teachers tried for twenty minutes to get him loose, then took him up to the kitchen. We heard rumors, never confirmed, that they had put his head in the oven and turned it up high. I contended at the time that this couldn't be true or his head would have exploded.

"The kid was fine, but he would never talk about what happened. What got me was not the event but the mystery afterwards. I still don't know how to get dry ice the size of a basketball off someone's tongue.

"I learned then that sometimes events themselves are easily explained but their repercussions impossible to divine. This loss of innocence soured me on science fairs forever."

Joe Versus the Volcano does not have a happy ending. But

Dianne Armer of Berkeley, California, believes a science fair steered her toward a career as a marketer of computer software—and taught her that lava is a many-splendored thing. "I entered my first science fair in sixth grade, and won second prize for building an electric motor out of nails, wire, and a dry cell battery," she recalled. "That was good. But the best part came at the PTA meeting that night, where kids weren't allowed and the nuns demonstrated all the science projects for the parents. Sister Vincent Mary put more chemicals than she should have into a volcano, and it kind of exploded and then oozed uncontrollably. We couldn't use the classroom for a week. I made my parents tell me the details over and over. I couldn't wait for the next science fair."

Sister Vincent Mary proved to be the last of the red-hot magmas. "To my great disappointment, the nuns didn't blow up anything at the next Parents' Night. But by then my interest in science was clinched."

When I asked Tina (not her real name), a computer programmer from Denver, if she'd ever entered a science fair and how she felt about the experience, she noted proudly that in fifth grade she'd made (everybody!) a volcano. "Just dirt on plywood. Mine wasn't a performing volcano. It just sat there, but I was very pleased with it. It's still in my parents' attic with my relief map of Scandinavia."

Dr. Karin Rabe, the first woman to teach in Yale's physics department, reminisced happily about her only science fair experience, as a senior at the Bronx Science High School in New York City. She talked about the excitement and pleasure of coming up with an original hypothesis—involving the effect of light on an enzyme that controls seed germination—even though, she insists, the work was sloppy. "I don't remember if I got a prize, but I do remember that one of the judges told me afterwards that it wasn't so much the project they were impressed with but the fact that I had read a lot and could answer questions specific to the project." She noted that the key to science fair success is doing a real experiment—formulating a hypothesis, testing it experimentally—and not just making a chart or a model. "But you know," she added, without prompting, "I remember a volcano demonstration that really impressed me. Someone basically made a pile of chemicals, lit them, and blue sparks come pouring out of the top as the middle caved in."

Jim Anderson, a designer at Pentagram Studios in New York, credits a fifth-grade science fair project at the East Hill Christian School in Pensacola, Florida with helping him learn creative problem solving. The problem: How to get white mice drunk. "I wanted to show how alcohol affects reactions and judgment. My dad helped me build an elaborate maze out of plywood. I ran some mice through, timed them, and averaged their time sober. Then I got them wasted.

"They wouldn't drink Scotch from a water bottle, so I had the idea of filling an eyedropper with Scotch and dropping it on their backs; they licked it off to clean themselves. That's the thing I'm most proud of. My experiment upheld other researchers' findings: In small amounts, alcohol is a stimulant, but in large amounts it's a depressant and a . . . confusant. The average maze time sober was three and a half minutes, and one mouse did it in half that slightly loaded. But one of the really drunk mice took half an hour.

"I won first place; mine was the most scientific experiment. About three weeks after this big party I threw them, one of the mice gave birth."

Twenty years later, Anderson is still plenty tickled by the story, and it's a pleasure to see. Then he had to spoil it all by adding, "The year before, I'd built a volcano."

The editor in chief of *Discover* magazine very nearly started life as a defrocked valedictorian because of a volcanic mishap. Paul Hoffman was a senior at Staples High in Westport, Connecticut, when he demonstrated his customized Fujiyama for a class of tenth graders. "I wanted to put on a really good show," recounted Hoffman. "I put in about six times the explosives necessary. Hot ash moved along the floor of the chemistry room like a lava flow out of Kilauea. Two girls in the front row wearing sandals started shrieking and ran out. I was banned from the chemistry lab and almost barred from being valedictorian, just for burning their feet."

Sometimes science fairs teach us things besides science and showmanship. Dennis Andersen, a textbook editor from Portland, Oregon, learned perseverance. "My fourth-grade project boiled down to a bowl of water, a fork, and a razor blade," Andersen remembered. "As people walked by, I demonstrated how surface tension kept the razor blade afloat when it was on its side but made it sink to the bottom when it was on its edge. I'd dip the fork in water, pick up the razor blade on the fork, and hold it out to passersby, saying, 'Would you care to try it?' One man, undoubtedly attempting to make a joke, said 'No, thank you, I don't eat razor blades.' I was mortified; I suddenly felt my science project was a sham and asked my parents if we could go home.

"But I sprang back, and the next year I conditioned a goldfish: I got it to come to the top of the bowl when I tapped a spoon, even when I didn't feed it."

Sometimes science fairs teach us to whimper and go belly up. Tina (not her real name), a computer programmer from Denver, told me that in sixth grade she spent weeks building a model of the solar system. But it didn't quite fit into the family car and when her dad tried to shut the door, Pluto bounced down the driveway, into the street, and under the tires of a pickup.

Recalls Tina (not his real name), a computer programmer from Denver, "In fourth grade, I made a sort of barometer that involved a soda straw and a hair. I was so proud of it! Then I got to the fair and this kid had done a thing about probability, a big production with continuously falling marbles. I felt stupid and miserable and knew right then that science wasn't going to be my thing."

It's the Tinas (not their real names) that many educators worry about. The Westinghouse Talent Search Winner off designing an experiment that resolves a centuries-old controversy over whether pterosaurs were bipedal can take care of herself.

(I was amused and touched by advice offered in Fort's book from the head of the Westinghouse Search, E. G. Sherburne: "Research takes time. You can't start a winning project at Thanksgiving and hope to win in December." For many of us, more realistic advice would have been: "You can't start a winning project during 'The Brady Bunch' and expect to be finished by Johnny Carson!" Synchronistic news flash: My sister just told me that a memorable "Brady Bunch" episode revolved around a volcano that refused to erupt on cue.)

Jerome Karle, winner of the Nobel prize in chemistry, told me that he fears the competitive aspect of science fairs may do more harm than good. "We all need encouragement, but especially children," he says. "I question how encouraging an atmosphere there is when large numbers of children make what they think is a considerable effort and only two or three are selected for the attention."

I decided to ask a modern child how he felt. Jarrod Fischer, ten, a fourth grader at the Redwood Heights Elementary School in Oakland, California, is a science fair veteran who has taken a blue ribbon in four out of five outings. "People at my school are still talking about my kindergarten project," he modestly asserts. Entrants are required to come up with a question that their project answers; his back then was How much weight can a snail pull? To find out, he harnessed those little suckers to a teensy Lego cart and loaded it with various objects—washers, a penny, some matches. This year, the question was What liquid makes broccoli grow the best? "I tried Coke, milk, beer, water, fertilizer, vinegar, cooking oil, bleach, and four more I can't remember. The best turned out to be fertilizer and water—or was it Coke? The worst was oil; the leaves started at one and a half inches and shriveled to a quarter of an inch over a month and a half."

Jarrod says he has lots of fun, but winning is important to him. "I think competition is okay, but I think each grade should be judged separately," he says. "This year I got a second place, and I was a little disappointed in myself."

A winning project, he explains, needs a good question and a good answer. "You also want to be a little funny, but not get too kookie. In my ice cube one [What liquid freezes fastest?], I had funny liquids like pickle juice and paint. We were thinking of using urine but decided not to." Much too kookie. "And we named our snails and our cars." (One year his question was Does the shape of a car affect its speed? "We stuck wheels on things and rolled them down a hill—plastic tubes, plastic cones, even a kosher salami. That never once made it down." "We" includes his parents, who encourage and suggest but don't do the work.) "We were thinking of naming the broccoli, but I did a funny drawing instead.

"Oh, and another quality of a winning science project is not making the question What happens to celery sticks when you put them in red water? I'd be disappointed if at least five people in our school didn't do that one and never get a ribbon. You'd think after a while kids would get smart."

Jarrod's pleased to report that brother Zachary, a kindergartner, is picking up the torch; he got a first place for asking the question Which liquid melts ice fastest? "I filled some cups with different things—regular water, dishwashing soap, salt water, and an egg," Zachary explains. "Then we put an ice cube in each cup and I had a stopwatch." Did he have to sit there and time the melting? "No," Zachary says. "My dad had to sit there. The funnest part was putting the stuff in the cups and mixing it."

I think the laureates Fort interviewed might agree that's the funnest part, even the ones against science competitions. Those include Nobel prize–winning geneticists Barbara McClintock, who believes that such contests thwart independent and original research, and Joshua Lederberg, who says he has a hunch that runners-up will make better scientists than winners, whose projects are too often all glitz. "There can be real creativity in failures," he told Fort. Linus Pauling also wrote to Fort that he was skeptical of the value of competitions.

Nobelist in chemistry Glenn Seaborg, on the other hand, called science fairs "exciting, constructive, and stimulating." E. O. Wilson wrote that he was glad his school had some kind of competition in which ". . . (unlike athletics) I could excel." And James Watson observed, "The important thing is to be able to do science while you're young." According to Fort, science fair booster Donald Knuth, a Medal of Science winner, thinks all sorts of competitions encourage creativity. "In seventh grade, a bunch of us spontaneously challenged each other at trying to diagram English sentences grammatically; that experience was later to help me in computer programming."

(I understand such grammar challenges between rival youth gangs the Crips and the Bloods can grow so tense that MIT linguist Noam Chomsky is sometimes called in to mediate.)

Don't listen to me; I'm just bitter because I didn't get to play with snails when I was a kid. All of us have good science in us, even the Tinas (not their real names). Few people realize that this was the real theme of Lerner and Loewe's famed musical, originally called *My Science Fair Lady*; Henry Higgins picks up flower girl Eliza Doolittle to prove to a crony that he can teach even a Cockney guttersnipe to build a model volcano.

Moonlighting

□ ■ □

The street vendor on Sixth Avenue in New York's Greenwich Village is selling something better even than a Dove Bar. He's hawking the secrets of the universe.

On the kind of September evening that's not yet sharply creased but still falls in soft, summery folds, Tony Hoffmann, in his early thirties, stands in the middle of the sidewalk beside a large white telescope with a hand-lettered sign taped to the tripod: VILLAGE ASTRONOMER—EXPAND YOUR HORIZONS. It's decorated with the stars and crescent moons favored by cartoon wizards in conical hats. "See Jupiter and its four brightest moons, only a quarter!" Who could resist such a bargain? Passersby line up for a look.

Hoffman, an ardent amateur astronomer and unpublished science fiction writer, took a day job not long ago as production manager for a publisher of computer magazines. But for more than a year, he's been making a good part of his living by revealing the hidden face of the cosmos to pedestrians. Tonight he answers their questions, earnest and facetious, with gravity.

"How far is Jupiter from Sixth Avenue?" asks a man who stops eating a tuna sandwich to pay his quarter and bend to the eyepiece of the telescope.

"Four hundred million miles," Hoffman tells him.

"At two bits a look, how much is that a block?" The Village

Astronomer isn't sure, but the man doesn't seem to mind. "Very beautiful! Thank you! Definitely worth the money!"

"What am I looking at?" asks a lovely young woman in a sweatshirt. Hoffman obligingly informs her, "The big ball is Jupiter; the two little balls on either side are its moons. Left to right you can see Callisto, Io, Europa, and Ganymede." "Wow!" says the woman with sincerity. Wow is right. What looks to the naked eye like an ordinary dot of light in an empty patch of sky is revealed to be a faintly mottled, luminous sphere set between two pairs of smaller, paler pearls.

Next in line is a boy about six. "What's it like up there?" he wants to know. "It's rather cool," Hoffman tells him. "Many astronomers think Jupiter is the most likely place in the solar system to find extraterrestrial life, since it has many of the same chemicals—hydrogen, methane, ammonia—that the earth's primitive atmosphere had."

The explanation is a couple of sizes too big for the boy, but this seems only to please him. "How do we get there?" he asks his father as they walk away.

Between questions from customers, Hoffman tells how he came to launch this enterprise.

"I got my first telescope when I was fourteen, and I became totally involved in observation," he says. "Later, at the University of Michigan, I studied astronomy, but I lacked the mathematical intuition it would have taken for me to succeed; I got a degree in creative writing." That meant clerical and sales jobs, mostly at bookstores. Although he continued to read about the heavens, he abandoned regular observations.

"Then, about a year ago, I happened to read a book called *London Labour and the London Poor,* by Henry Mayhew. It described occupations in London during the mid- to late 1800s, and one of these was 'street telescope exhibitor.' I was fascinated.

"A few months after that, I began writing a novel about a

future age when the world is all city and no one can see the stars; the science of astronomy has died. A fellow walks into an antiques store and comes across a telescope in a back room. He doesn't even know what it is but feels compelled to buy it."

Remembering the Mayhew book, Hoffman made a note in his journal that perhaps his fictional character should become a street astronomer. Then he thought that maybe *he* should. "I really became sort of obsessed with the idea," he recalls. "I hadn't had a working telescope for years, and I decided to buy one, even though I couldn't really afford it. I finally found a display model for $350."

He explains that it's a Selsi 3-inch refracting telescope ("refracting" means that the image is focused by a lens rather than by a mirror; 3 inches is the lens diameter). The tube is roughly 44 inches long, and the whole thing weighs about 40 pounds.

"In the beginning, I was afraid of several things," Hoffman continues. "That this was a frivolous investment, that someone might steal the telescope, that people would be nasty to me. The only trouble I've actually had is that people can be belligerently skeptical." One kid insisted there was no such thing as Jupiter. Several people have accused him of having a picture taped to the inside of the telescope. "These comments are made mostly in jest, but you'd be amazed how many people don't know that you can see a planet close up from New York City."

Tonight many of those who ante up stay awhile to kibitz, to exchange information and misinformation.

"Yo, what's that?" asks a tough-looking man wearing lots of muscles. He points to a star hanging over Chelsea. "That's Deneb, one of the most distant stars we can see with the naked eye," Hoffman tells him. "It burns with the light of 60,000 suns, light traveling toward us since the time of the Roman Empire, about 1,500 years." The man alleges he can dig it.

Three girls in their early teens, affecting world-weariness, ask, "What's this supposed to be?" Soon they're squealing and pushing each other aside for a look.

"I saw you here the other day, and I've been thinking about you," says a blond man wearing zipper–intensive clothing. "I thought it would be very cinematic to have this in a movie: People leave a restaurant and there's Jupiter."

Six guys with Guy written all over them, drinking Lowenbrau, ask if Hoffman is watching a girl taking a shower. (Of the fifty or so customers this night, eight make peeping tom jokes, five kid about Martians, and three reminisce at length about childhood telescopes. Two friends run into each other and say, "Small world!" Six people without quarters get freebies, among them a skinny kid who says, "I only got twenty cents. Can I just see three moons?" One hundred percent of Hoffman's customers say, "Thank you.")

A Brazilian couple, especially excited because they just visited the planetarium, stay half an hour, the husband translating Hoffman's explanations for his wife. (Customers tonight have included speakers of Portuguese, French, Italian, Spanish, and Japanese.) The Brazilian tells the six guys, "I saw four moons, but you're going to see eight. It's what astronomers call the beer effect."

Other people say:

"This is really great! Jupiter!"

"That's a streetlight, not Jupit—oh. Oh!"

"I read that people actually live on the sun; it's only a magnetic inversion that makes it look like it's on fire."

"I've never looked through a telescope before. This means a lot to me."

"What are those stripes?" (They're cloud banks on the planet.) "How much is this magnified?" (Sixty times.) "How do you know it's a planet?" (Stars are twinkling points of light; planets are glowing spheres.)

The Most Enthusiastic Viewer Award goes to a fourteen-year-old boy so excited by the telescope that he looks as if he's going to go supernova, his arms and legs spiraling off to opposite ends of the galaxy. Hoffman tells him, "New York astronomers have a lot to contend with. There's air pollution and excess light—about 90 percent of the light of the stars is washed out—hot air rising off roofs, lack of open space, crowds."

But the crowd is the best part, even more thrilling than the sight of five far-off shining balls in the sky. Hoffman's customers radiate something you don't often see on the streets of New York: a pure form of pleasure, joy in its gaseous state before it coalesces into hard, dark chunks of disappointment and resentment. Tonight, people who in another context could be frightening or annoying are exhibiting a touching willingness to be instructed. They're innocently absorbed, like an infant tasting her feet for the first time. These people are impressed. These people are satisfied. These people are a small, temporary community eager to pass on their freshly acquired knowledge to others in line when Hoffman is busy, glad to be looking at things much larger and more permanent than themselves. These people are getting a big bang out of the universe.

When Jupiter goes behind a building, Hoffman moves to Seventh Avenue between West Tenth and West Fourth streets. He changes his pitch. "See the lunar plains and craters, only twenty-five cents! Satisfaction guaranteed, or your money back!" Through the telescope, the moon is so close you feel you could get there in fifteen minutes. People are even crazier about the hometown satellite than they were about Jupiter.

The last viewer of the night, a man from Mexico named Jorge, is so delighted with the moon that Hoffman finds him a double cluster of stars in the constellation Perseus, about 8,000 light-years away. Says Jorge, before he runs to catch a bus, "I look at the sky and I feel better. I see one star, and I look at the world different."

Shortly after midnight Hoffman closes shop. "The other night I was showing a family Jupiter," he says as he folds the tripod, shoulders the telescope, and heads for home, a few blocks away. "I said, in response to some question, 'This is reality.' The little kid asked, 'Mommy, what's reality?' And the mother answered something like, 'It's what's not in the movies.' So much of a New Yorker's exposure to things is secondhand—photographs, movies, plays, even the news. When people look through a telescope, they're looking at something that's real. We have so few chances to look at the thing itself. So what I'm doing changes people's lives.

"I'm making money; business is moving a lot better than I'd expected, especially since I can't work when it's cloudy. But my sense of mission overrides any other concern. I feel a strong kinship with the other London street astronomer I read about. If I ever finish my novel, I'll dedicate it to him."

About the Author

Judith Stone is a contributing editor at *Discover* and *Glamour,* and was formerly articles editor at *McCall's* and *Science Digest.* "Light Elements," her monthly humor column for *Discover,* won a 1989 National Headliner Award. Her work has appeared in *The Village Voice, Newsday, Elle, Vogue, Ms., Travel and Leisure* and many other publications. She was a member of the touring company of Second City, the improvisational theatre troupe, and she knows what DNA is, although she recently had to be told, very gently, that the yolk of an egg doesn't harden into a chick. She lives in New York.